Handbook of Aroma Research

MANFRED ROTHE

Introduction to Aroma Research

Handbook of Aroma Research

Collection of monographs written by international flavour specialists

Edited by
MANFRED ROTHE
Potsdam-Rehbrücke

Introduction to Aroma Research

By

MANFRED ROTHE
Department of Aroma Research, Central Institute of Nutrition, Academy of Sciences, GDR, Potsdam-Rehbrücke

First English Edition

With 71 Figures and 35 Tables

Kluwer Academic Publishers

Dordrecht / Boston / London

Library of Congress Cataloging-in-Publication Data

CIP

Rothe, Manfred.
Introduction to aroma research.

(Handbook of Aroma Research)
Translation of: Einführung in die Aromaforschung.
Bibliography: p.
Includes indexes.
1. Food—Analysis. 2. Food—Odor. I. Title.
II. Series
TX535.R6713 1986 664'.07 86-17819
ISBN-13:978-94-010-7135-2 e-ISBN-13:978-94-009-1419-3
DOI: 10.1007/978-94-009-1419-3

Distributors for the U.S.A. and Canada
Kluwer Academic Publishers,
101 Philip Drive, Norwell, MA 02061, U.S.A.

Distributors for Albania, Bulgaria, Chinese People's Republic, Cuba, Czechoslovakia, German Democratic Republic, Hungary, Korean People's Republic, Mongolia, Poland, Rumania, the U.S.S.R., Vietnam, and Yugoslavia
Akademie Verlag Berlin

Distributors for all remaining countries
Kluwer Academic Publishers Group,
P.O. Box 322, 3300 AH Dordrecht, The Netherlands

Published by Akademie-Verlag Berlin, DDR-1086 Berlin, Leipziger Straße 3—4, in co-publication with Kluwer Academic Publishers, P.O. Box 17, 3300 AA Dordrecht, The Netherlands

Kluwer Academic Publishers incorporates the publishing programmes of D. Reidel, Martinus Nijhoff, Dr. W. Junk and MTP Press.

Original title: **Einführung in die Aromaforschung**

Preface to the first English edition

In 1976 when the first German manuscript of this "Introduction" was completed aroma research was described as being in a phase of rapid development. This new scientific discipline had spread, from its origin as a specialized part of food chemistry, over a broad and differentiated area between physiology of flavour sensation, aroma analysis — with a fascinating spectrum of instrumental and sensory techniques — to such practical fields as food technology and flavouring production. In the meantime this tendency has continued. In many aspects our knowledge has now reached a higher level, and we can recognize many more connections between different aroma problems. It has therefore been necessary to complement the original material by new results, and to summarize some aspects in order to give the most up-to-date picture possible of our discipline.

One problem, however, remained unchanged: An introduction has to present a review of the field under discussion without going into too much detail. A detailed discussion should principally be confined to the special monographs of this series. The author had to select a few appropriate examples out of a large number of possibilities, which is really a rather difficult task. Another aspect to be considered is the necessity of demonstrating the actual facts in a readily understandable way for the non-specialist who may be entering the field for the first time. Furthermore there was a demand for a translation into the English language. In spite of a concept that the individual monographs of this series were to be written either in English or in German language we feel that some volumes should be available in both languages. Moreover, we had to consider that, in the main, activities in the flavour field have developed within English speaking countries. Consequently, most authors and experts have published their results in English.

I want to thank the Akademie-Verlag Berlin and especially lector Karl ABEL for an always good and effective collaboration. All our problems were discussed and solved in a fruitful and cooperative way stimulated by an open and friendly atmosphere. Within the Central Institute for Nutrition Potsdam-Rehbrücke I am indebted to our director, Prof. H. SCHMANDKE, as well as to Prof. H. RUTTLOFF for their understanding and help. Finally I should like to thank Mrs. Edith WEISE for typing the manuscript and for assistance in the English translation.

Potsdam-Rehbrücke
Manfred ROTHE

Contents

1. Research in the flavour field — present situation and trends

1.1. Human perception of odour and taste

All living beings perceive their environment by sensory contacts registered from the surroundings, and by comparison of these impressions with similar ones kept and stored in memory. The sum of such effects provides a concept of the environment. A continuous increase in such experiences results in an ever improved adaptation to living conditions, and to an optimized pattern of life behaviour and life circumstances.

The contribution of the individual sense organs to our world picture depends on the number of stimuli as well as on their quality. Sensitivity and effectiveness of the various sense organs are increased in the course of generations, the more they are used and stressed. The stage of development of living beings indirectly characterizes their ability to register and to integrate stimuli of various types and intensities and to compare the result with earlier information stored. It is this type of comparison which causes conclusion and experience and is finally reflected by behaviour.

If we compare human senses of odour and taste perception with the other sense organs, we cannot overlook the fact that they are less well adapted than our senses of sight and hearing. In modern industrial society, we are continuously exposed to optical and acoustical stimuli. Speech, traffic, industrial noise, displays from technical equipment, music, theatre, radio or television, they can all be considered as examples. In contrast to this frequent use of our physical senses the 'chemical senses'[1]) are called into use only periodically at special times fixed, for example, at meal times.

Comparing the human senses of smell and taste with those of other higher organisms, we find many examples of higher sensitivity (dog, fish, butterfly and other insects), but few with more effectiveness. In the case of the silkworm moth **Bombyx mori**, a few molecules of the sex attractant hexadeca-10*trans*-12*cis*-dienol (bombycol) are sufficient to produce an irritation signal. Only this one special chemical, however, can produce an effect with such a high sensitivity. Very small changes in the structure decrease the odour effect by orders of magnitude (D. SCHNEIDER 1971, see also table 20).

The efficacy of the human senses of smell and taste is characterized not only by high sensitivity but also by a very good ability to discriminate between the various nuances of flavour qualities. Normal adults can distinguish between nearly 2000 odour impressions. There is, however, a certain indolence in our chemical senses, because of their normally low claim to recognition and definition of known odour effects. Thus the qualification for participating in a sensory test panel can be achieved or improved by training. A trained panelist in the flavour industry discriminates between up to 10000 odour impressions (K. B. DOVING 1963). Flavour perception is influenced by interactions between our different sense organs. Frequently there is some difficulty in differentiating between odour and taste as two separate physiological senses. Of these two senses, odour perception is much more complicated. In contrast to the ability of adults to discriminate between much more than 1000 single aroma impressions, there are only four basic taste qualities.

[1]) Odour and taste are stimulated by chemical stimuli, the other senses by physical stimuli

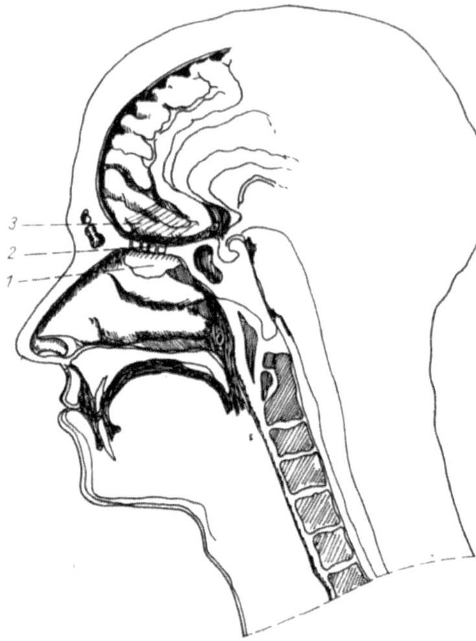

Fig. 1
Olfactory organs of man
1 = regio olfactoria
2 = cribriform plate
3 = olfactory bulb

Odour perception takes place in the upper part of the nasal cavity only, the so-called **regio olfactoria** (olfactory sensory epithelium). This part of the nasal mucosal is only 10 cm^2 in area, but it contains some 10 million receptors. There are three principle cell types:

1. **Olfactory receptors** which detect, decode and transmit the sensory information about quality and intensity of odour
2. **Sustentacular cells** which add gland-produced mucopolysaccharides to the mucus layer on the epithelial surface
3. **Basal cells** which seem to be stem cells becoming active in the course of normal cell turn-over (T. V. GETCHELL and M. L. GETCHELL 1977).

There are two possible ways for odorous substances to reach the **regio olfactoria**. One way is the transport of volatiles together with the breath air stream, via the nasal cavity to the loungs and vice versa. Another way is via the nasopharynx connecting the mouth with the nasal cavity. This is what happens during the chewing and swallowing process in the mouth, when food aroma compounds are partially volatilized after coming into contact with saliva and slightly increased temperature. In both cases volatilized molecules of aroma components pass the **regio olfactoria**.

After sorption in the mucus layer over the active surface the stimulus sets off an electrical signal within the smelling cells which are fitted with six to eight micro hairs per cell. This signal is conducted via the **cribriform plate** to the so-called **bulbus olfactorius** (olfactory bulb) in the front brain. The further way of the information to the **odour field** of the brain is unknown neither its registration nor the method of storage, or the comparison with impressions received and stored earlier is understood: in fact we do not completely understand even the most elementary aspects of receptor function (D. G. MOULTON 1971).

The receptors for **taste** on the tongue surface register the four basic qualities **sweet, sour, salty** and **bitter**. The perception is located in special areas of this sense organ (see figure 2). Anyone can confirm this phenomenon by careful observation.

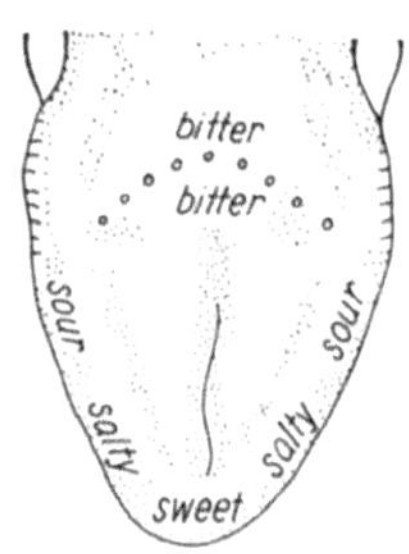

Fig. 2
Sensitivity of tongue areas for the basic tastes
(R. W. MONCRIEFF 1967)

Each of the mushroom-shaped papillae on the tips and edges of the tongue bears 8 to 10 taste buds; each of the circumvallate papillae at the base of the tongue has 100 to 300. Each bud lies beneath a pore, and is composed of 40 to 60 spindle-shaped taste cells packed together like segments of an orange. So-called microvilli are extendend into the pit of the pore. Here the interaction between stimulus and receptor is assumed to occur (D. G. MOULTON 1982).

Besides the four main principles mentioned, a few more effects are discussed as taste sensations. G. LILJESTRAND and Y. ZOTTERMAN

(1956) added the perception of **alkaline taste**, R. W. MONCRIEFF (1967) the **metallic taste** to the four main principles. Another typical taste effect is the so-called **Umami** (Y. KAWAMURA 1987) observed in connection with the consumption of meat. It is discussed as an effect of substances called **flavour potentiators**. Monosodium glutamate and 5′-ribonucleotides belong to this group (see chapter **1.3.3.**).

Apart from the histological structure of its elements the physiology of odour and taste reception has not yet been elucidated nevertheless several odour and taste theories have been published. They are all based on presumed connections between the molecular structure and the shape of special odour or taste molecules and their aroma and/or flavour impression. The principle was discussed firstly by L. PAULING (1946). Examples are the **Stereochemical Theory** (J. E. AMOORE 1952, 1971, 1982; J. E. AMOORE and D. VENSTROM 1966), the **Vibrational Theory** (R. H. WRIGHT 1954, 1969) and die **Adsorption Theory** (M. G. J. BEETS 1957, 1982; E. T. THEIMER and J. T. DAVIES 1967; P. LAFFORT 1969). So far, however, the application and validity of all these theories is limited to a relative small number of flavour substances only. Moreover, none of these theories can give a clear answer to the question of what molecular or functional properties are responsible for the inclusion of a substance among the aroma of taste compounds (W. STEINER 1973).

There have, however, been various attempts to elucidate the complex field of odour and taste perception. Only a little of this very interesting research can be mentioned here. One of the new concepts, for example, is the electrophysiological study of odour perception by insects. These animals are equipped with receptors of an extremely high sensitivity for some volatile compounds and thus have been used successfully for test purposes. Many interesting results have been published in this field within the last three decades. Of special interest is not only the high sensitivity of the antennae of male insects, which can register the female over distances of kilometers, but also a very high selectivity and specificity (D. SCHNEIDER 1969, 1971; W. A. KAFKA 1971; see also table 19 and 20, p. 64).

Another current research area uses the determination of electrophysiological signals in connection with similarity studies of odour qualities. Electro olfactograms from frogs and rabbits as test animals indicate a number of high correlation effects between homologues of odorous substances or between pairs of odour stimuli (K. B. DOVING 1966; P. MACLEOD 1971).

Starting-points for future activities in the complicated field of reception mechanism can also be found in publications about sweet and bitter sensations. S. SHALLENBERGER and T. E. ACREE (1967) claim the presence of two functional groups within the molecule of a sweet substance which can form hydrogen bonds. They must be separated from one another by a particular distance. These attributes are claimed to be a common characteristic of many sweetening agents. Thiocarbonyl compounds equipped with neighbouring π-systems represent one example of similar relationships for the bitter taste (R. MAYER and F. WITTIG 1972). For bitter peptides, interactions between hydrophobic side chains in the molecule and the receptor protein are discussed (A. PETRITSCHEK et al. 1972). The intensity of the bitter taste after hydrolysis can be calculated if the amino acid composition of the protein is known (K. H. NEY 1972). A recent review of H.-D. BELITZ and H. WIESER (1985) summarizes our knowledge on structure — activity relationship in the field of bitter compounds. From the tongue of beef and pork proteins have been isolated which possess binding capacity for sweet or bitter substances (F. R. DASTOLI et al. 1968).

So far, the activity of physiological research has dealt only with the first step, the reception of chemically produced sensory impressions. The subsequent processes may be still more complicated. Our ability to perceive and to integrate a very complex mixture of flavour compounds in less than a second, and to compare this sensation with earlier stored impressions, still remains 'a mystery' (A. A. ALBRECHT and L. M. BIEDLER 1966). However, in recent years our knowledge in this field has begun to increase. There are some preliminary ideas about the sequence of molecular, membranous and neural events responsible for the transmission of the reception signals to the brain (T. B. GETCHELL and M. L. GETCHELL 1977).

1.2. Attributes contributing to flavour sensation

As already mentioned flavour sensation results from a cooperative effect of odour sensation and taste sensation. Whereas taste substances like sugar, sodium chloride, acids or bitter compounds are registered during the mouth passage by the taste receptors, volatile odorous components reach the sensory olfactory region directly via the nose or indirectly via the connection between mouth and nose cavity. In the case of solid foods, the chewing and grinding process, together with the influence of temperature and saliva, causes a partial volatilization which allows entry through the pharynx cavity to the nasal area.

In English as well as in German the combined odour/taste sensation is often described simply as **taste (Geschmack)**, although usually the total sensation is under discussion. This was one of the reasons for introducing the term flavour. **Flavour** describes the overall sensation and that means something more than **odour** and **taste** alone. In figure 3 which demonstrates the relationship between these terms *haptic* or *tactile* feelings are included additionally as a third group of sensations. The concerted action of these three groups of sensations constitutes the **flavour**.

As shown in figure 3 there are many cross-links. What we define as the typical flavour of a food in most cases is mainly identical with pleasant odour (= aroma) perceptions. Thus **aroma** is central to total flavour perception. Products absolutely lacking in aroma will never be characterized as having 'good flavour'. Aroma, however, is frequently pleasant only if we feel a simultaneous taste sensation. A well-known example is the importance of a special sugar/acid ratio for the flavour quality of fruits or of wines. Canned fruits without added sugar seem to be low in flavour so do fully-ripe oranges, apples or pears which may have sufficient aroma components and a sweet taste, but no acid. Salt-reduced bread, sausages produced without addition of sodium chloride or beer without bitter substances are further examples of tasteless and insipid flavour. In all these cases the content of volatile aroma components may be just as high as normal.

Besides odour and taste, further sensations contribute to the overall flavour impression in most foods and dishes. Among these consistency effects are the most important. Many examples indicate our ability to integrate consistency impressions, such as hardness, toughness, stickiness, elasticity, plasticity or viscosity, into flavour.

Texture and consistency of food and meals produce physical impressions like touching, pressing, resistance against grinding, or, in the

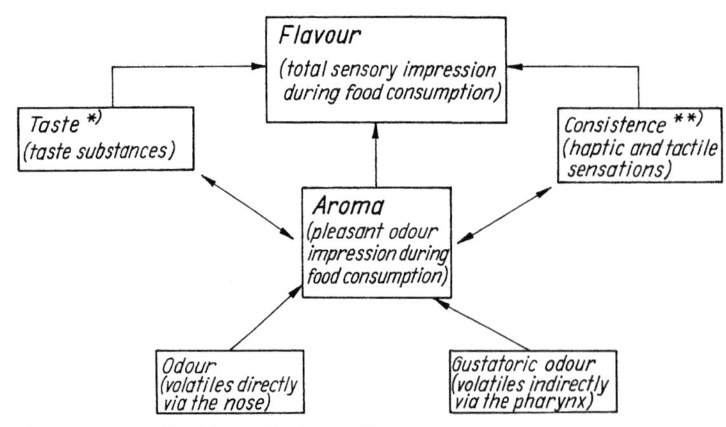

Fig. 3
Relationship between different senses in 'flavour' and 'aroma' perception

case of beverages, a liquid, viscous, smooth or pudding-like feeling. The effect of such sensations in parallel with the aroma and taste sensation is well-known: We are accustomed to consuming crackers, crisp bread, toast, potato chips or similar modern snack which have a dry, splintering consistency. If such products have a changed consistency caused by increased water content we feel an empty, abnormal flavour, and will usually refuse to eat the product. If such foods vary only in their moisture content they must, however, have the same aroma and taste as before. This fact demonstrates the influence of consistency on the overall flavour impression.

Texture feelings are also included into the flavour evaluation produced during the chewing of steaks, the bite into a slice of pineapple or cucumber. If we imagine eating these products in a homogenized mashed form, it is obvious that the overall flavour impression would decrease.

Although appearance and colour are not included into the term **flavour** they represent further factors influencing food acceptance. The influence of colour on flavour identification in flavoured jelly sweets is demonstrated in table 1. All sweets were prepared with the same raspberry aroma. The colour was the only difference.

Liqueurs coloured with tasteless blue dyes are available on the market, however, they are sold in rather small amounts only. Many consumers refuse to buy them. They find such products unnatural as blue colours are rather seldom associated with food items. The same is true for colours which are generally not observed in particular foods, like green-coloured milk or red-coloured bread. Normal sensory panels reject these products, though they have an acceptable flavour which is not in the least changed by the colouring. In tests with the same group a normal flavour evaluation resulted when the same coloured products were offered in the dark.

The sensory attributes discussed are not independent, but may overlap and influence each other. According to F. J. FRANCIS (1977) colour is a very important sensory factor because of its quick and safe registration. G. URBANYI (1984) proved this observation on the examples of potatoes and strawberries with different flavour, colour and texture. The samples were offered to the test panel three times: in normal form, after mashing for excluding texture influences, and after colouring for avoiding the help of colour as a differentiation tool. Especially in the case of small differences, flavour evaluation was influenced by the colour (compare table 2).

Table 1
Identification of a raspberry flavouring in jelly sweets of different colour
(J. SCHEIDE 1975; recognition of the same aroma in coloured sweets by 18 people)

Colour	Green	Blue	Yellow	Red
Right answers	2	3	4	8
Wrong answers influenced by the colour	5	—[1]	11	5
Wrong answers without association	11	15	3	5

[1] no association because normally blue colour in jelly sweet is lacking

Table 2
Influence of the colour on the perceived aroma of foods
(Extract from work of C. M. CHRISTENSEN 1983; % increase in the sensory score compared with the uncoloured sample)

	Aroma intensity (coloured/ uncoloured)	Aroma quality (coloured/ uncoloured)
Cheese	+49 %	+43 %
Margarine	+67 %	+48 %
Orange drink	+55 %	+58 %

With regard to some foods, temperature and acoustic sensations can be included into flavour impression, too. The optimum flavour reception of white wine, champaign or beer, for example, is produced at a somewhat lower temperature than the flavour of red or dessert wines. Coffee or soup seem to decrease in flavour quality if their temperature is reduced below the normal hot state. The same is true for warm beer, or ice cream in the case of temperatures increased some-

what above zero. Such optimum temperatures for flavour perception, however, depend on custom and may be learned by experience (J. F. CAUL 1970).
Chocolate, or roasted peanuts, are especially pleasant for many consumers if they cause a definite noise while being crunched in the mouth. The same is true for biting into a whole apple, or for chewing of crisp bread or potatoe chips which must be crunchy.
C. M. CHRISTENSEN and Z. M. VICKERS (1981) found the crispness of such products to be highly correlated with loudness. There was, however, no difficulty for the test panel in determining crispness when the sounds produced by chewing were blocked by a masking noise. Vibritional feeling caused by fracturing crisp food is considered to be the reason.
These examples indicate possible interactions between the different types of sensation. An optimal flavour can obviously be observed if several attributes participate in the production of total perception. In such examples of combination only one possibility is absolutely lacking, i.e. the registration of an optimum effect without any aroma reception. Taste substances alone seem to produce a one-sided impression. This again proves the central importance of aroma within flavour perception.
E. von SYDOW (1971) discussed the order of stimulation during eating, and its importance for food acceptability. First of all, the consumer

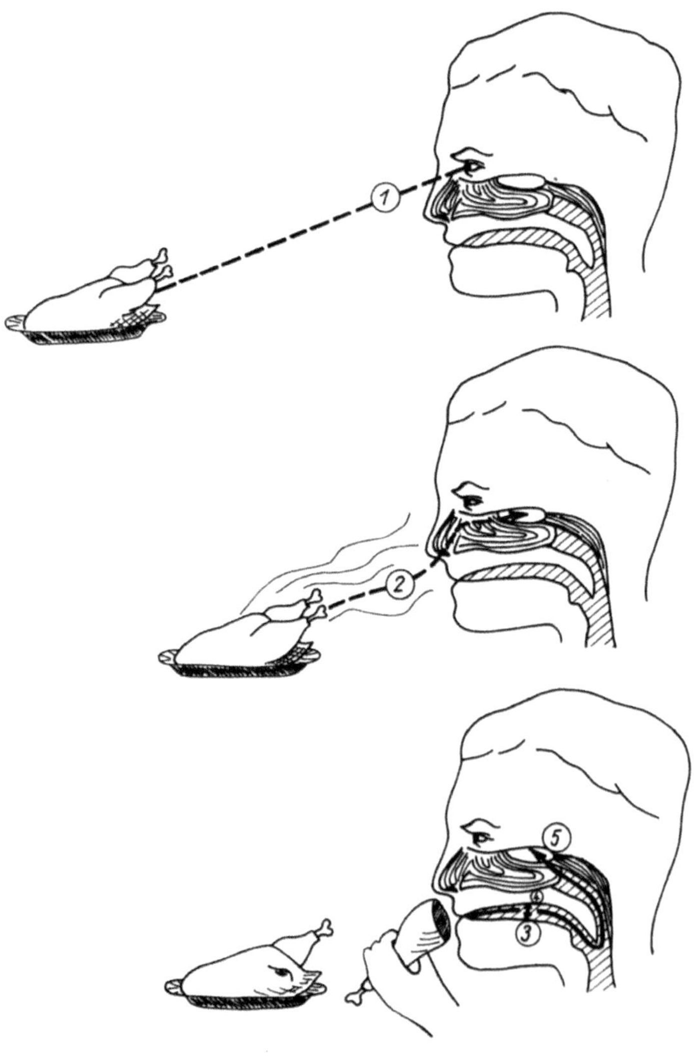

Fig. 4
Order of sensations during eating
1 = optical sensation
2 = odour resp. aroma sensation
3 = taste sensation
4 = consistence or texture sensation
5 = odour resp. aroma sensation

is confronted with the appearance and colour of the food or meal; later on with its odour or aroma, and finally, during the chewing process, with consistency, taste and aroma once more. Figure 4 demonstrates this ranking principle.

The temporal order, however, influences the total flavour impression. Independent of the real importance and weighting, the initial sensations have a predominating influence. Considering in addition that we take in as much as 87% of our information from the surroundings by sight, 9% by hearing and only 4% by the other senses (G. T. ZAUSCH 1971), the overwhelming importance of the appearance of foods and meals becomes evident. To a high degree our food selection depends on this. There is nothing new about this observation: Even in ancient times the effect of a good presentation of foods and meals was well-known and was used to stimulate the appetite. This is true for the arrangement of dishes, as well as for the use of red- and yellow-coloured fruits or green vegetables which improve the total appearance of the meal by means of colour contrasts. On the other hand unusual colour effects may induce a negative opinion.

1.3. Terminology

In the past there has been confusion about the use of the main terms in flavour research. This was influenced by temporal change in the meaning of certain words, as well as by a different interpretation of terms in various languages. But for a clear understanding among the specialists it is necessary 'to speak the same language'. Thus there are efforts to standardize the most important terms on a national and international basis (compare H. FINCKE 1949; R. L. HALL 1968; H. GLATZEL 1968; G. T. ZAUSCH 1969, 1970, 1975; ANON. 1970, 1975; ISO 1978; H. MAARSE 1981; L. HALL and E. J. MERWIN 1981, D. G. MOULTON 1982).

There is no doubt that, of all terms internationally used in the flavour field, the word **aroma** is the oldest. The term originates from ancient Greek, and was in early times identified with the odour sensation produced by spices. Later on, all pleasant odour sensations were summarized under this term, such as those produced for example by fragrant herbs, spices and other seasonings. The homeland of most spices is the Asian continent. Due to its early high culture and widespread sea trade, Greece was the first of the European countries to come into contact with the aromatic herbs and spices imported from Asia (W. TREIBS 1956).

The term **aroma** is known in most languages, but has not always been applied with the same meaning. While orginally used for seasonings and their effect, the term later came to be associated with essential oils or their ingredients. Now it has been extended in such a way that it is synonymous with the total pleasant odour perception produced during food consumption. Thus at present *aroma* is used for all characteristic and pleasant food odours, no matter whether they are intensive or at a low level. We now discuss not only the aroma of coffee, of chocolate, or of citrus concentrates, but also the aroma of bread, of milk, or of vegetables, which are characteristic, but low in aroma intensity (D. J. TILGNER 1961).

As shown graphically in figure 3 the term **flavour** describes the total sensory impression including aroma, taste and consistency sensations. Only the appearance is excluded from all sensory impressions. *Aroma* holds the center within the sensation; it is responsible for all typical flavour impressions. Thus it is justifiable to use the term *aroma* instead of *flavour* in many cases. This implies that terms such as *flavour research* and *aroma research*, *flavour concentrate* and *aroma concentrate* or *flavouring* and *aromatization* may be used synonymously.

H. G. MAIER (1970) claimed that nearly all chemical substances of low molecular weight have a certain odour or taste effectiveness. However, their odour impression is generally not pleasant and in most cases has no relation to the aroma of foods.

According to figure 3 **taste substances** are defined as those components which produce taste effects within the mouth during chewing or drinking (R. RIKLIN 1967). Thus the application of this term should be restricted to sweet concentrates based on sugars or sweeteners, and to acids, sodium chloride and bitter substances. Only such compounds can be related to the effect of *taste substances* acting within the mouth cavity and influencing the taste impression directly. Monosodium glutamate and carbon dioxide represent the best-known examples.

Although the definition of the three terms **flavour, aroma** and **taste** seems to be clear, the real situation is much more complicated. This is due not only to languages but also to the fact that different groups are interested in flavour problems: e.g. the flavour industry, the food industry and nutritional scientists. Up to now, however, there have been no general standardizations of the terminology.

The flavour industry in several countries, for example, often summarizes flavourings under the term **taste substances (Geschmackstoffe)**. This definition is not correct, because nearly all concentrated flavourings contain volatile odorous compounds as dominating ingredients. The use of such terms, however, is caused by the fact that this industrial branch produces flavourings for the food industry as well as numerous odorous fragrances for the cosmetics industry. They must therefore use different terms to distinguish.

The application of flavourings in very different concentrations represents another problem of terminology within the flavour industry. There is a variety of different names in each language, e.g. *essences*, *aroma bases*, *basic flavourings* or *aroma concentrates*. It might aid understanding if only the term **aroma concentrate** were used to describe the whole group. This definition is justified by the fact that the main attribute of this group of additives consists in the rather high concentration which necessitates their application in small amounts.

The term **flavourings** is used for all types of natural, nature-identical and synthetic mixtures for flavouring purposes. It is a blanket term summarizing all groups of aroma and flavour intensive additives used within the food industry.

1.3.2. Off-flavour

After the definition of the word **flavour** and its introduction into other languages, the term **off-flavour** should also be defined as an international standardized term. **Off-flavour** comprises all those forms of deviation from normal food flavour quality which produce an abnormal flavour impression. Off-flavour can originate from single substances which are normally not present in the food in question. Furthermore it is possible to register odorous substances as off-flavour by changing the normal sensory profile to a one-sided strengthening of special notes. In recent years J. M. H. BEMELMANS and M. C. TEN NOEVER DE BRAUW (1975), M. J. SAXBY (1982) and F. B. WHITFIELD (1984) have reviewed different types and origins of food off-flavours. Figure 5 tries to summarize them.

Rancidity, bitterness and soapiness, strong acidity and mouldiness belong to those types of off-flavour which have long been recognized. Today they are accompanied by numerous types of fermentation off-flavours caused by unwanted microbial metabolic reactions. This group includes ester-like, diacetyl-like or sulphurous off-flavour.

Further modern off-flavour sources are some pollution effects of our advanced civilization. Pesticides or herbicides in uncontrolled high doses, for example, sometimes cause strange flavour notes. Chlorophenol impurities can produce especially strong effects. This class of chemical substances must be discussed not only as the direct origin of strong (medicinal) off-flavour, but simultaneously as the most potent precursor of this type of food deterioration. Sometimes mysterious off-flavours occurring in warehouses, stores and food processing plants observed in recent years have been related to the presence of chlorophenols in fungicides, wood preservatives and agrichemicals. Specific and detailed research in this field has elucidated that a microbial transformation of chlorophenols into

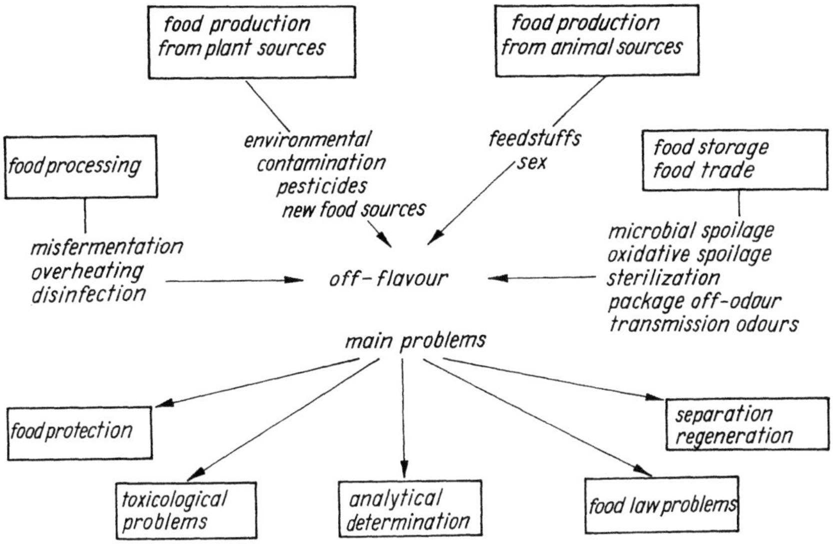

Fig. 5
Origins for off-flavour in foods and main problems
(see also J. M. H. Bemelmans and M. C. ten Noever de Brauw 1975; M. J. Saxby 1982)

the corresponding chloroanisoles is important because these substances belong to the most effective odour compounds known. Thus for 2.4.6.-trichloroanisol a concentration of 3×10^{-14} (in water) was found as a threshold value, while in wine concentrations of 2×10^{-12}, i.e. 0.002 µg/l could be detected as an off-flavour (H.-R. Buser et al. 1982).

Chlorinated phenols themselves can also cause flavour defects directly. They are reported to be formed in meat processing plants by a spontaneous reaction of phenol traces with heavily chlorinated water used for disinfection purposes. In some cases plastic fittings containing traces of free phenols in contact with chlorinated water have spoiled beverages passing through them after the water cleaning disinfection step. Some of the chlorinated phenols, such as 6-chloro-2-methylphenol and 2.3.4.6-tetrachlorophenol have been found to develop off-flavour in the concentration range of only microgram amounts per kilogram.

This high flavour activity of the chlorophenols and chloroanisols easily causes the contamination of foods via the air. Such cases have been found in the neighbourhood of chemical plants which produce these substances and transmit off-flavour by air pollution. The easy permeation of such types of off-flavours through packaging materials is a related problem (see also C. Engel et al. 1966, N. M. Griffiths and D. G. Land 1973, R. F. Curtis et al. 1974, F. B. Whitfield 1983, H. Maarse 1985).

Production and processing of new food sources represents a further problem of modern food industries which is often said to be due to off-flavour. Plant protein concentrates and isolates, which are of increasing interest from the economic point of view as well as for developing countries, may be mentioned. An early example of such a product was soya protein, which appeared on the U. S.-market 15 to 20 years ago. This by-product of soya processing contains bean-like off-flavour substances which can also be found in other leguminose proteins. The off-flavour is transferred to the protein-enriched product and may inhibit the consumption of these foods. Although some of the contaminating compounds have been identified in the meantime (geosmin; R. G. Buttery 1976, see figure 6) problems of economic separation or masking still remain. Effects of flavour changes can also be observed when proteins of plant or milk origin are added to cereal foods in order to improve their nutritive value. A high purification degree of such protein concentrates

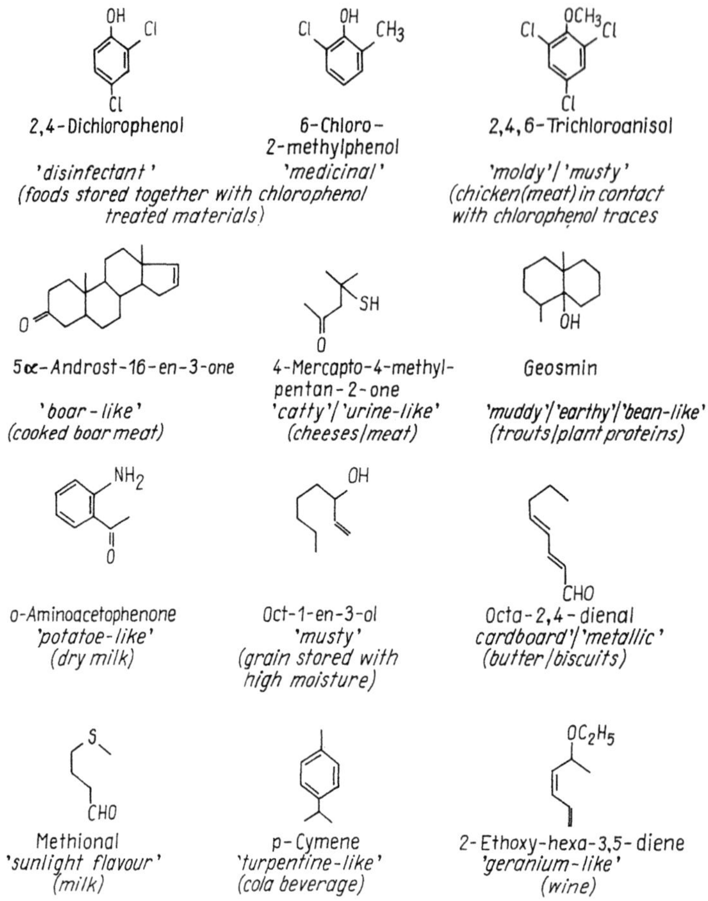

Fig. 6
Off-flavour producing substances identified in special foods

is the best solution of the mentioned problem but it is often too expensive.

Figure 6 shows some types of off-flavour substances recently discussed in the literature and summarizes some of these problems. Special types of off-flavour can also result from a change of the normal spectrum of aroma components to render it more one-sided. A typical example is diacetyl, which produces flavour changes if it is present in raised amounts in beer (L. Nykänen and H. Suomalainen 1983) or in wine (H. H. Dittrich and E. Kerner 1964) or ropy wheaten bread (M. Rothe 1963). Extremely high amounts of acetaldehyde, hydrogen sulphide, methane thiol, dimethyl sulphide and methylbutanal are typical for some kinds of off-flavour in milk (H. T. Badings and C. de Jong 1984).

Another example of off-flavour formation is an increased amount of free fatty acids in cheese. Whereas normally the rancid and soapy notes produced by acids of medium chain length (C_4 up to C_{10} especially) show a positive effect on total flavour sensation, the same components may be responsible for an off-flavour in higher concentrations. According to figure 7 there is a positive contribution of rancid and sharp sensory notes to flavour in the range of low intensities with high correlation coefficients up to an intensity of approximately 2.5 or 3.5 on the 5-point-scale used. Concentrations above the turning point show a continuously decreasing effect on total flavour quality (M. Rothe et al. 1982). The same is true for some foods with bitter attributes like beer, vermouth or chocolate, which cause a negative impression if they contain too high concentrations of bitter-tasting compounds.

There are a few additional examples of this type of off-flavour caused by a disturbed

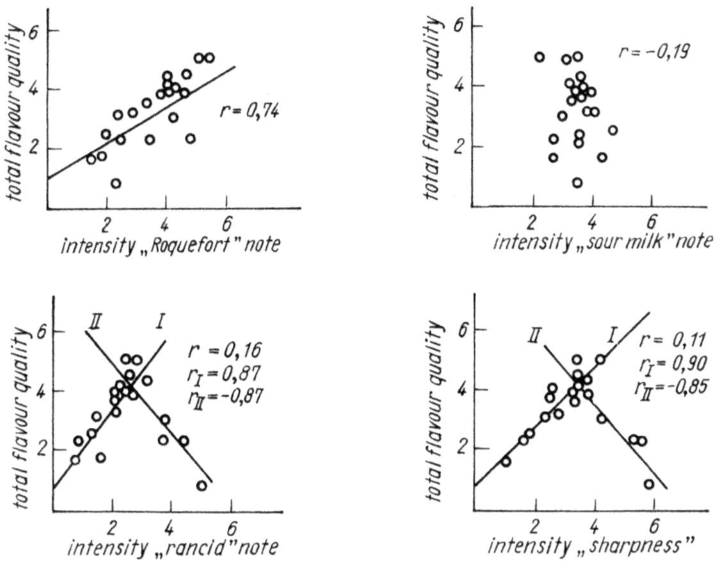

Fig. 7
Correlation between total flavour quality of Blue cheese and single sensory attributes (Comparison test of 20 samples with different quality; sensory 5-point scale for each attribute)

ratio of aroma compounds. In apple juice, for example, a very high content of *cis*-hex-2-enal or of fruit esters produces negative flavour effects, though these components are normal ingredients of this beverage. In the case of sauerkraut an extremely high amount of short chain fatty acids (C_3 to C_6) causes 'cheesy' off-flavour (M. L. VORBECK et al. 1961). T. PERSSON and E. VON SYDOW (1973) and T. PERSSON et al. (1973) found a deviating ratio of carbonyl to sulphur components to be correlated with the characteristic off-flavour of canned meat products.

On account of the high sensitivity of milk and milk products to develop off-flavour, this medium is especially appropriate for a discussion of the different sources of contamination. Based upon 146 literature items, W. F. SHIPE et al. (1978) published an excellent review of nomenclature and standards, and a bibliography of off-flavours occurring in milk products. They differentiate between the types indicated in table 3.

As shown in figure 6, off-flavours can be produced by physical means like heating or irradiation, by chemical or biochemical reactions like lipolysis, or by special types of microbially or oxidatively induced changes of the aroma spectrum. Last, but not least, direct or indirect transmission odours like odourous feedstuffs or weed odours (silage, fish meal, green forages, wild garlic) or off-flavours

Table 3
Categories of off-flavours in milk
(W. F. SHIPE et al. 1978)

Causes	Descriptive or associative terms
Heated	cooked, caramelized, scorched
Light-induced	light, sunlight, activated
Lipolyzed	rancid, butyric, bitter, goaty
Microbial	acid, bitter, fruity, malty, putrid, unclean
Oxidized	papery, cardboard, metallic, oily, fishy
Transmitted	feed, weed, cowy, barny
Miscellaneous	absorbed, astringent, bitter, chalky, chemical, flat, foreign, lacks freshness, salty

absorbed from the environment, may play a role. In milk products a kind of off-flavour often observed is a bitter taste which originates from peptides produced by an uncomplete protein hydrolysis.

Considerable difficulties are connected with analytical significant proof of the volatile food contaminants present. Figure 8 demonstrates this using recently published results of F. B. WHITFIELD (1983). The chromatogram contains only trace amounts of chloroanisols. It is obviously very problematic to find these traces out of more than 100 accompanying volatile compounds. Here the method of splitting the gas stream at the bottom of the gas chromatograph, followed by sniffing the

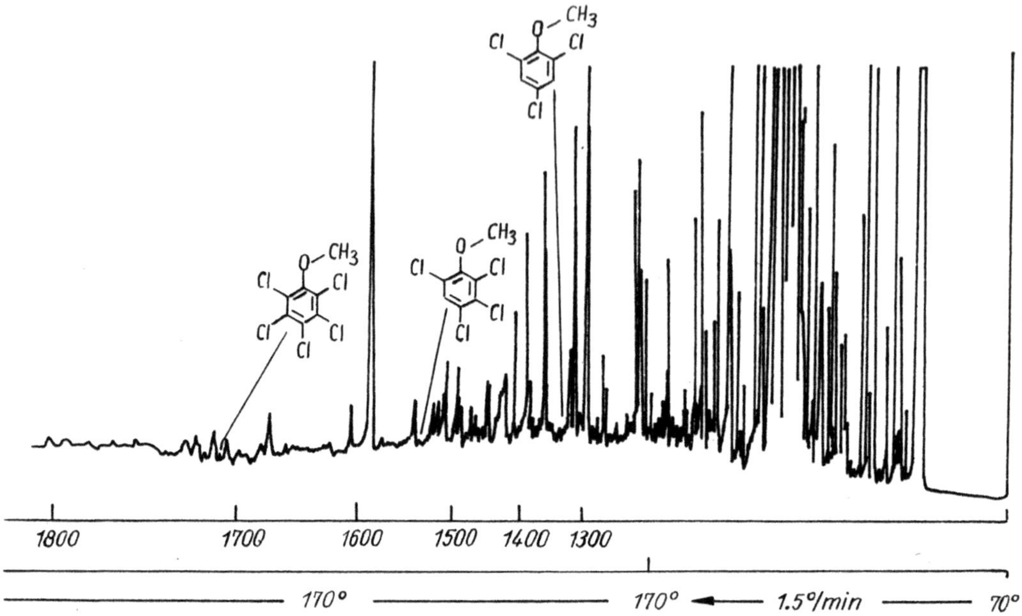

Fig. 8
Gas chromatogram of volatiles isolated from dried fruit with mouldy off-flavour
(F. B. WHITFIELD 1982)

carrier gas leaving the equipment, was applied successfully (see also figure 33 on p. 52).

1.3.3. Flavour potentiation and modification

Flavour potentiator or the synonymous term *flavour enhancer* defines a special group of flavour active components contributing to total flavour sensation indirectly, by their presence only[1]. Flavour potentiators have no own flavour activity in themselves but are able to enhance the aroma or taste of special flavourings (A. KUNINAKA et al. 1978, J. SOLMS 1967). They occur in several foods naturally, and were first identified by K. IKEDA in Japanese seaweed in 1912. Besides some typical East Asian foods, such as fermented soya sauce, meat products, cheeses and mushrooms are the main foods in modern nutrition which contain these substances in higher amounts. Consumption rates are summarized in table 4; a review was recently published by J. A. MAGA (1983).

[1]) New results discuss this effect as a fifth basic taste (Y. KAWAMURA 1987).

Within recent decades, flavour potentiators have developed into one of the most important groups of flavouring additives. Among them monosodium glutamate and 5'-ribonucleotides are the additives with by far the highest world production rate. Using data of A. KUNINAKA (1978) and J. A. MAGA (1983), the distribution of the world production of monosodium glutamate in the main producing countries can be estimated as in figure 9. The differences between some East Asian countries and other parts of the world depend on traditional consumption habits originating from the typical rice diet in that region which is very poor in flavour and requires added flavouring (see also table 4). Monosodium glutamate and 5'-ribonucleotides improve and enhance the typical flavour of meat-containing products like canned meat, soups and snacks, as well as the flavour of some vegetables and milk products. Physiological tests on cats and rats as test animals prove that this effect is due to a stimulation of the taste receptors within the mouth (M. SATO et al. 1965, A. ADACHI et al. 1965). An increase in sensibility of the taste receptors to sodium chloride is significant, but this obviously does not represent the only effect.

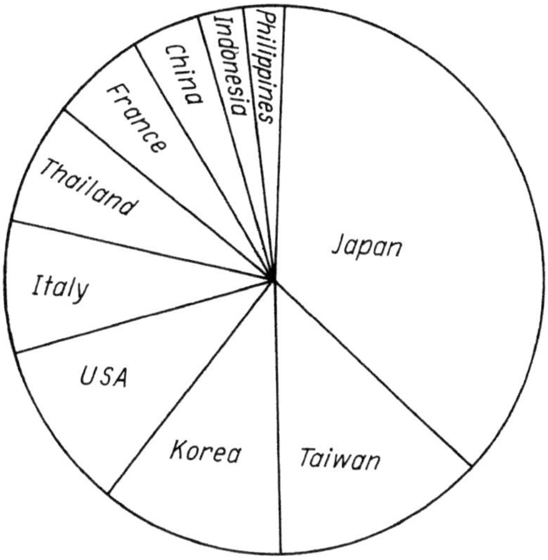

Fig. 9
Distribution of the world production in monosodium glutamate to the major participating countries (% of total production; compare A. KUNINAKA 1978; J. A. MAGA 1983)

Table 4
Consumption rates for added monosodium glutamate as flavour potentiator in different countries
(T. GIOCOMETTI 1979 and own data)

Country	Consumption (g per caput and day)
Taiwan	3.0
Korea	2.3
Japan	1.6
Italy	0.4
USA	0.35
GDR	0.05

It is supposed that the flavour potentiation may be considered as an effect similar to that of carbon dioxide in aqueous solution, which stimulates the sensibility of the taste receptors to sour and salty sensations (Y. KAWAMURA and A. ADACHI 1967). Comparing the quality of tap water or boiled water on one hand with that of fresh spring water or mineral water on the other hand consequently leads us to prefer the more harmonized sensation of the latter carbon dioxide containing waters.

If flavour potentiators influence taste receptors, the chosen term is right. Thus direct effects on taste contribute to total flavour sensation. On the other hand, ribonucleotides also have *flavour modification* properties (M. H. WOSKOW 1969). In protein hydrolysates they provide a masking effect against such off-flavours as *hydrolysate-like*, *sulphury* or *burnt cabbage*. Here we seem to be on the borderline between a real flavour potentation and synergistic effects between aroma and taste components.

Such interactions may be responsible for the creation of additional terms. K. H. NEY (1971), for example, discusses *aroma intensifying effects*. Such new terms, however, cannot be clearly defined and delimited as long as their physiological base is not understood.

A few examples of real interactions between odour and taste substances may easily be found. In fruit juices saccharose releases an aroma potentiating effect which cannot be related to physiological reasons only, but may depend on physical factors. Parallel aroma and taste sensations may play a role, too. The same is true for the aroma enhancing effect of sodium chloride in salt-free meat products or in bread. Similar to the masking activity of 5'-ribonucleotides in meat products, saccharose is able to mask negative odour in

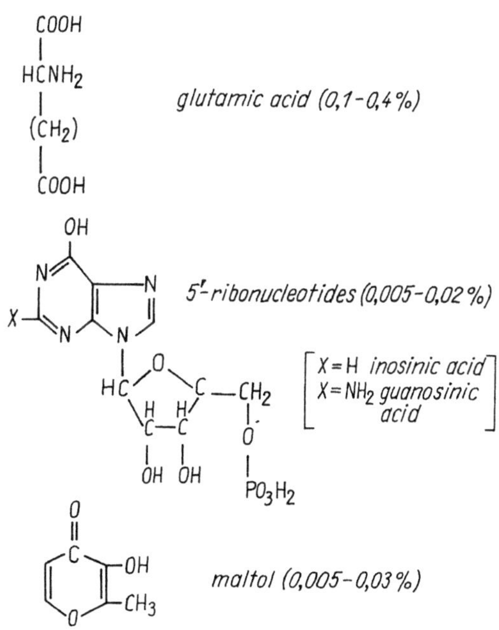

Fig. 10
Flavour potentiators used as food additives

processed fruits. E. von SYDOW et al. (1974) observed this for *grassy* or *sharp* notes in fruit juices. Aroma-enhancing activities, however, are typical for other sweeteners, too. Such effects are discussed in the patent literature (M. R. CLONINGER and R. E. BALDWIN 1974).

Are we therefore free to define sweeteners or salt as *aroma enhancers*? Without any physiological proof this question cannot be answered clearly. In an indirect way physical effects will probably also contribute. Saccharose, for example, changes the volatilization rate of aroma components in very high concentrations (W. W. NAWAR 1966, A. G. WIENTJES 1968).

In figure 10 maltol (2-methyl-3-hydroxypyron) is included as a flavour potentiator for sweet sensations. But, despite the definition, this frequently used flavouring has a typical malt-like aroma. Thus some scepticism as to the consistent usage of the term **flavour potentiator** may be justified. As discussed in the literature on beverages, 5 to 15% of the normal sugar content can be replaced by maltol (R. D. OLSEN 1964). Possibly this additive acts indirectly by giving sweets or beverages a new aroma nuance, dependent on the sugar concentration.

There are, however, some examples of real **taste modification** in literature. The sweetening after effect of chewing artichokes is an old example. The bouquet of wine used as a table beverage may be spoiled by this flavour change. Most people, but not all, are sensitive to this effect, which lasts for approximately five minutes. The substances responsible are chlorogenic acid (3-caffeoylquinic acid) and cynarin (1.5-dicaffeoylquinic acid) (L. M. BARTOSHUK et al. 1972).

Basic glycoproteins within the tropical *miracle fruit* (*Synsepalum dulcificum*) are genuine **taste modifyers**. When the ground-up fruit comes into contact with the mucous layer in the mouth, sour components are registered as sweet ones (D. E. INGLETT et al. 1965, K. KURIHARA and L. M. BEIDLER 1968, E. WIYAND et al. 1970).

Taste perception can be inhibited, too. Thus the sweet sensation is diminished after chewing the leaves of the tropical plant *Gymnema sylvestre*, which is used as a tea plant in India. The effect lasts for hours, and operates with sugars as well as with sweeteners or sweet-tasting foods: sugar crystals in the mouth feel like sand. The receptors for sweet stimuli in the taste buds are blocked by acetyl glucuronides of gymnemic acid, a pentacyclic hexahydroxytriterpene (W. STÖCKLIN et al. 1967, K. KURIHARA 1969).

The example of flavour potentiators demonstrates the difficulty of arriving at a clear definition. As long as the physiological base of such effects is not known with certainty, it is not easy to differentiate between terms like **interaction, synergism, potentiation** or **enhancement.** An international vocabulary would help to change this situation.

1.4. Aroma and flavour: the consumer's point of view

Food quality is a complex, not standardized term, its definition depends on different points of view. In the case of food, however, quality can be discussed as the sum of several physical and chemical attributes, changing from food to food in their type and importance. **Physical factors** can contribute to total quality directly, as do such attributes as appearance, colour, consistency, texture or viscosity. Examples of the indirect influence of physical factors on the total quality of foods are such functional characteristics as water binding, emulsifying or gelatinization capacity, or instantization.

The content of main nutrients including vitamins, mineral substances and fibers as well as protein quality or dietetic value, belong to the **chemical quality factors**. Aroma compounds, as another chemical factor, are of high importance, and are often differentiated from other quality attributes. From the consumer's point of view, flavour and aroma take the first place among the various quality factors of food.

Other groups, however, may discuss the quality problem in a different way. For the food producer in agriculture or the food industry economic and technological parameters are the most important ones. Raw materials which deliver a high yield, or which are characterized by a good and standardized processing behaviour, are considered by the producer to be those with the best qualities. A third interested group, the nutritional scientists, consider nutritional factors to be the most important. There are many interactions between these three groups, as shown in figure 11.

Consumers, however, are less interested in the origin of foods or in their nutritional value. For them flavour quality is by far the most important, although knowledge of nutritional problems is continuously broadened by nutritional education and propaganda. But, in general, nutritional recommendations are followed mainly by people whose health requires a special diet.

The relationship demonstrated in figure 11 is simplified. There are differences from food to food. It depends on the type of product whether, and to what extent, factors like colour, appearance, consistency or texture contribute to total food quality.

The high preference for flavour in consumer acceptance of different food qualities can be demonstrated statistically. J. J. POWERS and M. C. QUINLAN (1974), for example, tested correlations existing between acceptance and factors like appearance, colour and flavour in the case of canned peaches. According to table 5 flavour, with the highest correlation coefficient of 0,73, ranks in first place, followed by texture in second place.

Flavour has been responsible for food selection for centuries (see page 100ff.). A recent observation dates from the years after World War II. At this time, when undercaloric nutrition was widespread sometimes food or food substitutes of bad quality were refused or at least uncompletely digested (H. GLATZEL 1968). Human instinct for the selection of foods with special flavour or off-flavour seems to have triumphed from earliest times right up to the present day. Today food composition in different areas of the world is still

Table 5
Correlation between the acceptance of canned peaches and single sensory attributes
(J. J. POWERS and M. C. QUINLAN 1974, correlation coefficients)

	Colour	Appearance	Flavour	Texture
Acceptance	0.42	0.39	0.73	0.54
Colour	—	0.58	0.36	0.38
Appearance	—	—	0.25	0.37
Flavour	—	—	—	0.44

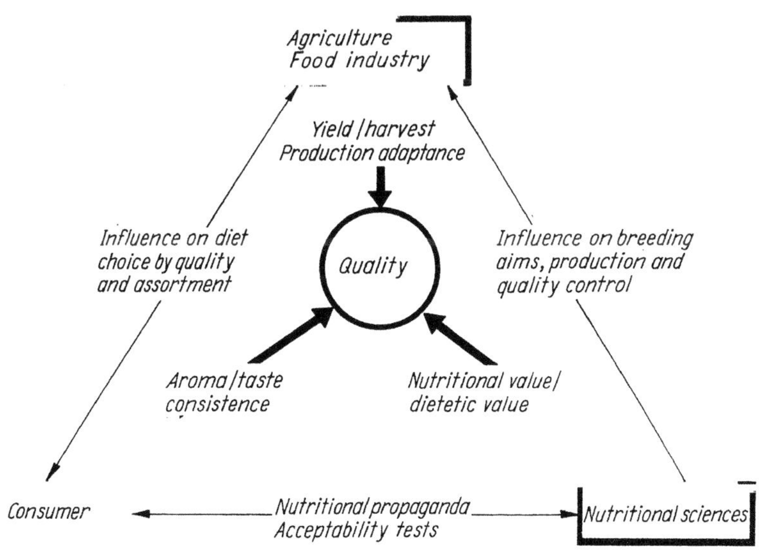

Fig. 11
Main factors for food quality in the view of different groups

Table 6
Examples for the percent weight of different food attributes within the total flavour evaluation standards of the GDR
in percent of the total value; according to R. NEUMANN 1983)

Food	Flavour evaluation	Chemical and physical factors	Hygienic factors	Nutritional factors
Rye bread, unpacked	90	—	—	10
Wheaten bread, packed	60	30	—	10
Wheaten flour	40	50	10	—
Pastry	50	30	10	—
Puddings	70	10	10	10
Soups, instant broth	80	—	10	10

influenced by traditional eating behaviour. There are considerable differences between the European or North American diet and such diets as are usual in African, Middle American or East Asian countries. Modern collaborative tests, however, dealing with sensory evaluation of flavour-rich foods all over the world, indicate that in principle food preferences and food acceptance are the same for all human beings (B. LUNDGREN et al. 1978, 1983).

In the present century there have been large changes in food consumption habits. These have been caused by the formation of new societies and the development of old ones and by rapid scientific and technical progress. Modern traffic conditions make distances seem small. Consequently, foods from all parts of the world are at our disposal everywhere. The process of equalizing food consumption and food flavour habits has made great progress this century. The tests mentioned above show clearly that, apart from traditional and geographical habits in food consumption, there are no obvious physiological differences between different races or continents. The successful expansion of the market for special flavour-rich foods like cola, citrus and passion fruit, tomato and bell-pepper, vanilla or margarine is therefore easy to explain.

The importance of aroma and flavour within total food quality is reflected in the food evaluation schemes which are used in many countries for quality control within factories or by official state organizations. In such evaluations aroma and taste dominate. Table 6 demonstrates this fact using some control schemes applied in the GDR as an example. Internationally there is a trend to increase the percentage aroma and taste taken from the total value, whereas in the past technical and physical attributes of food quality have been more to the fore. This trend should be supported if we consider the interests of the consumer.

1.5. Development of aroma research

1.5.1. Historical and modern aspects

Three decades ago aroma research was a domain of the flavour industry. **Aroma** was not only defined as a sensory perception, but was also a term used for aroma concentrates of natural or synthetic origin. Such products were applied for flavouring of sweets, ice cream, confectionary, beverages, canned products or margarine as well as for the seasoning of cooked dishes. The flavour industry responsible for the production of such concentrates has a long and interesting history. The use of spices and seasonings can be traced right back to the beginnings of human culture in Egypt and China, several thousand years ago.

The early preference for special nuances of flavour, however, may also be connected with centuries of hunger for much of the

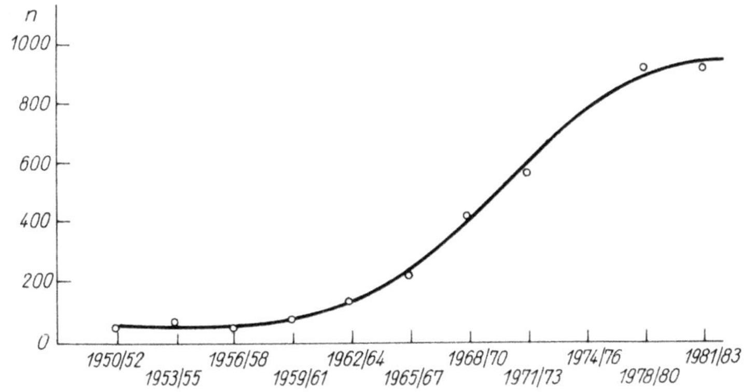

Fig. 12
Development of aroma research indicated by the number of publications
(Calculation based upon abstracts published in Z. Lebensmittel-Untersuch. u. -Forsch. and Documentation "Nahrung und Ernährung" of Central Institute for Nutrition Potsdam-Rehbrücke)

population in most countries. Food shortage led to the necessity of storing and preserving; by thermic processes like roasting, baking or frying, by salting or smoking, or by the use of special herbs and spices. The aroma and flavour effects produced in the course of such treatment gradually developed into a 'bonus' for the quality of foods (I. M. Wightwick 1973). People thus became accustomed to a special flavour and aroma which was considered to be the most important quality factor. In this way, procedures originally necessary for preservation of food were carried out later only for the production of special flavour qualities.

While hunger may have been the first reason for the discovery of flavour producing food processes, within the last century the technical revolution of food industries has brought about further progress. In this short introduction there is not enough space for a detailed discussion of this interesting historical development. However, some of the interesting facts are summarized in the historical synopsis on pp. 100—108.

During the last three decades flavour research has developed remarkably. As can be seen from figure 12 which is based on the number of publications on aroma problems in well-known abstracting journals[1]) the rate of activity has increased enormously.

After a rather constant rate up to 1955 an exponential rise can be observed. The start of this trend dates from the introduction of chromatographic techniques into aroma research. Encouraged by the progress in this field, interest in aroma problems has increased continuously since then (see for example F. Rijkens and H. Boelens 1975, H. E. Nursten 1979, H. Maarse 1985).

A similar trend can be observed on counting the number of aroma components identified in different years. This is possible on the basis of the lists published by S. van Straten und F. J. G. de Vrijer (1977) and H. Maarse and C. A. Visscher (1984, 1985). Figure 13 summarizes all volatile components which are said to occur in food products (H. Maarse 1985).

In figure 14 the development of our knowledge about aroma components of three particular foods is included, i. e. coffee, apples and bread (M. Rothe 1980).

The curves in figure 14 have a characteristic shape, showing a turning point. This profile may be influenced by the fact that, after a period of effective use in identification, the

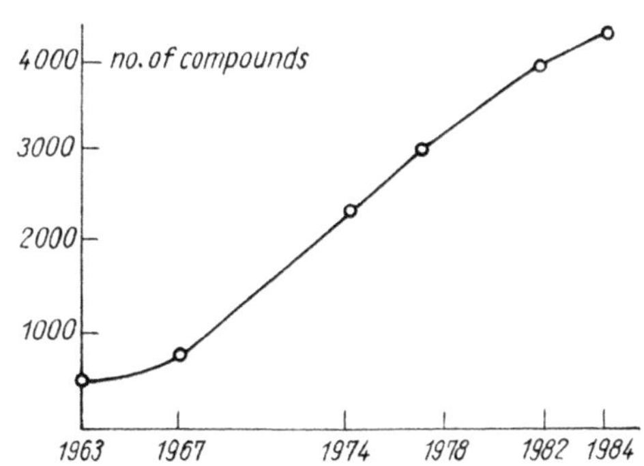

Fig. 13
Number of volatile compounds reported to occur in food products
(H. Maarse 1981)

[1]) Only those publications were considered where aroma problems are in the central of the discussion

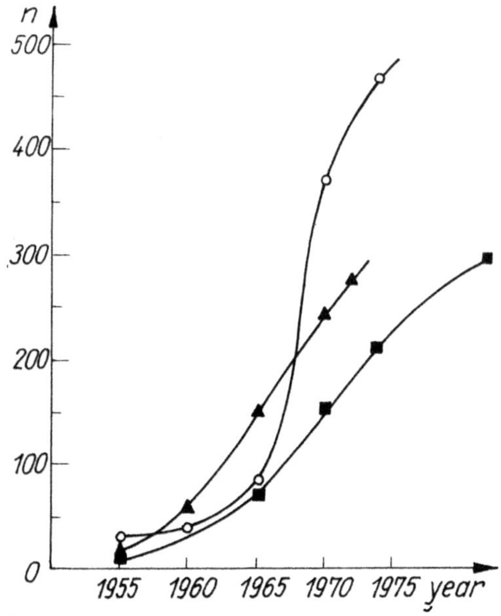

Fig. 14
Number of identified compounds in three foods
○ = coffee ▲ = apple ■ = bread

modern gas chromatographic and spectrometric techniques are now about to reach their limit. If no more new methods and principles for identification are discovered, the number of volatiles will soon attain its maximum. Following these trends, F. Rijkens and H. Boelens (1975) estimated a total number of aroma components as somewhere between 5.000 and 10.000; by 1985 nearly 4.500 have been listed.

The number of aroma components identified up to now depends not only upon the techniques available. There are differences between foods, in their aroma concentration as well as in the economic value of different kinds of foods, which may influence the research potential. Both facts may be responsible, for example, for the much higher amount of volatiles in coffee so far registered. Looking for the reasons for the increase of activity in the flavour field, we find there are several influences. As mentioned above, the introduction of gas chromatography into food analysis some years after the well-known initial publication of A. T. James and A. J. P. Martin (1952), is the most important one. These procedures first made it possible to separate and determine a complex of compounds in a concentration as low as 10^{-5} to 10^{-9}. In food, aroma-effective volatiles occur in such concentrations: only some spices and seasonings reach higher levels.

But it was not only the introduction of modern methods that increased interest in the aroma field. The application of these analytical tools caused a rapid development of separation and identification techniques, as well as their coupling with spectrometric methods. Important landmarks were:

— the use of capillary chromatographic columns (M. J. E. Golay 1958, G. Dijkstra and J. de Goey 1958),
— the introduction of head space techniques (R. G. Buttery and R. Teranishi 1961),
— the change from constant temperatures in gas chromatography to temperature programs (S. dal Nogare and C. E. Bennet 1958, W. E. Harris and H. W. Habgood 1966),
— the development of new detector systems with special selectivity (flame ionization detection; I. G. McWilliam and R. A. Dewar 1957, 1958; J. Harley et al. 1958; electron capture detection: J. E. Lovelock and S. R. Lipsky 1960; flame photometric detection: S. S. Brody and J. E. Chaney 1966),
— the use of organic polymers as stationary phase (O. L. Hollis 1966) and
— the coupling of gas chromatography with mass spectrometry (R. Ryhage 1964).

The use of these technical advances in analytical methods has opened the way to a jungle of flavour evaluation problems which previously could be only partially and insufficiently elucidated. Today we are aware that the aroma of most foods is caused by some hundreds of odorous compounds (see table 18, p. 59). But the more complex food aromas become, by the identification of new volatiles, the more we have to direct our activities to a complete identification of the aroma spectrum.

Another problem of great importance is the weighting of the numerous aroma components. Modern gas chromatographic techniques are an indispensable tool for solving this task. However, along with the improvement of modern equipment there are continuously increasing costs.

Table 7
Trends as to the amout of modern chromatographic and spectrometric methods applied in aroma research
(% of the tested number of publications taken from the documentation "Nahrung und Ernährung" of the Central Institute for Nutrition Potsdam-Rehbrücke)

Publication years	Tested number of aroma publications	Classical procedures[1] and reviews %	GLC and HPLC procedures %	GC/MS procedures %
1956/1960	300	95	4	1
1965/1969	1000	77	14	9
1970/1971	400	71	19	10
1972/1975	700	58	25	17
1981/1983	1000	52	30	18

[1] sensory, colorimetric and enzymatic methods, paper and thin layer chromatography

The rapid progress in flavour research due to gas chromatographic analysis can be confirmed in different ways. The data summarized in table 7 are based on the documentation 'Nahrung und Ernährung' of the Central Institute of Nutrition Potsdam-Rehbrücke. This documentation comprises the most important journals in the field of food and nutrition. The data in the table were again taken from relevant publications dealing directly with aroma problems.

The development and introduction of modern analytical methods, however, are not the only reason for the rapid progress in the flavour field during the last 30 years. We must also consider the technical revolution that has affected the technologies of all highly developed countries in this century. Modern processes for food production are more rationalized, simplified, shortened, highly mechanized or partially automated; they are accompanied in most cases by an unfavourable decrease in aroma intensity and/or change in aroma quality. The same effect can often be observed in the case of new convenience products, pre-cooked or otherwise instantized snacks or dietetic foods.

Some typical examples are given in table 8. They show only the principle involved: the effect discussed need not be valid in every case. There is no doubt that the economic situation of most countries makes it advantageous to shorten and simplify food production as much as possible, or to develop continuous techniques. Most of the fermentation and ripening processes involved in the production of alcoholic beverages, milk products or bread are traditionally very time- and space-consuming. Consequently there must be an interest in rationalizing. The same is valid for all thermic processes of drying, cooking, roasting, frying, baking or sterilizing. But a decrease in flavour quality frequently results. The solution of aroma problems connected with such rationalization processes is an important task for the food industry. It seems to be relatively easy to simulate the physical properties of a food produced under modern conditions. Often it is more difficult to achieve the traditional flavour in such foods. As an example for the large differences in time needed figure 15 demonstrates old and modern rye bread technologies (M. ROTHE and H. RUTTLOFF 1983).

But technical progress will not be inhibited by such aroma problems. We have early to look for ways of producing flavourful food in spite of modern production, processing or preparation techniques. Important steps in this direction are obviously a complete qualitative analysis of the aroma complex followed by quantification. This seems to be another problem responsible for the increasing interest in flavour research.

The extension of such flavour problems throughout the food industry represents another reason for more research in this area. Only a few decades ago, the flavour industry itself felt responsible for solving them. The term **flavour** was identified with flavouring

Table 8
Modern food technologies causing aroma losses or changes

Food	Changed or new production step	Technological or economic advantages	Aroma effects
milk	high temperature sterilizing	rationalized use of energy extended shelf-life	aroma changes
cheese, raw sausages	shortening the ripening process	reduction of storage capacity	aroma reduction
wine, high quality spiritis	shortening the ripening time	reduction of storage capacity	lacking storage bouquet
bread	shortening the fermention step	reduction of dough storage	reduced fermentation aroma
bread	shortening the baking process	rationalized use of energy	sharp and burnt crust aroma
chocolate	shortening the conching process	reduction of space and storage capacity	aroma reduction
coffee, tea	instantization	preparing, weight reduction in trade	aroma changes
pudding potato purée	production of prepared, gelatinized products	preparation	aroma changes
fruit juice concentrates	transformation into dry state	rationalized use of energy, weight reduction in trade	aroma reduction and aroma changes
potatoes	peeling by alkali	economizing working power, rationalized preparation	aroma losses and aroma changes

and flavour-producing problems. Up to that time aroma research had been limited to the production of sweets, confectionery, biscuits, beverages or margarine to give some examples of foods with no distinct aroma of their own.

More recently, however, the interest on flavour research has extended also to those types of food in which the flavour and aroma are produced in the course of processing. Though such foods account for the main part of total food production and food consumption, very little research activity was concentrated on their flavour before 1950. There was a

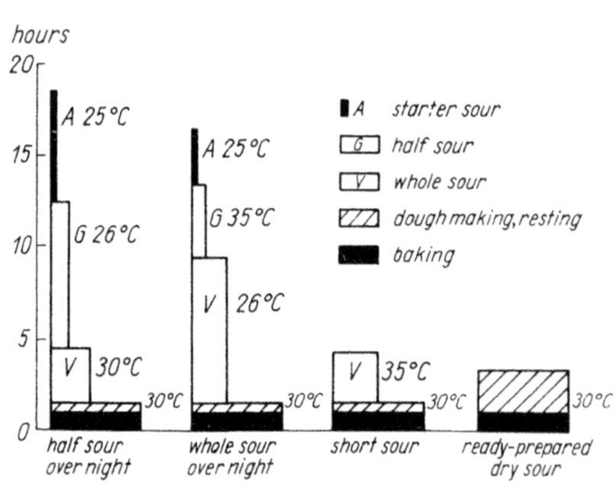

Fig. 15
Required time for sour preparation in different ways and its influence on flavour evaluation

gap in knowledge between the typical flavour problems of the flavour industry dealing with essential oils and some traditional flavourings on one side, and the flavour problems of the processing industry on the other. Since then this gap has been filled by many publications.

Another current problem of flavouring should be mentioned here. Progress in modern food production has enabled the food industry to produce more and more 'simulated' foods based on pure proteins and carbohydrates. Such isolated nutrients can be produced from new economic food sources: plant proteins from soya bean or other legumes are an example. In the future these proteins, as well as those of microbial origin, may well become significant not only in animal feeding but also in human nutrition. The production of plant (or microbial) protein is much more economic than that of animal protein, where allowance must be made for large losses caused by protein degradation for energy production in the animal. Considering the enormous increase in world population and the need for much more food, simulated products of the type mentioned seem to have an increased chance for success (M. V. Tracey 1969, R. J. Flannary 1975, B. Gassmann 1977).

But the trend to simulated foods is not only promoted by economic problems. The availability of new, highly purified, nutrient compounds provides opportunities to change the nutritional composition of foods in any way that may be required. Thus the food industry is increasingly able to produce so-called 'tailor-made' foods, which have a special field of application and a special market for dietetic purposes. As discussed in detail in chapter 3.3., the increasing malnutrition in highly developed countries is accompanied by many problems relevant to national health. Thus there is an increasing request on the market for more dietetic products.

There are, however, many flavour problems involved. According to table 9, a change in

Table 9
Flavour changes in dietetic foods caused by differences in the nutrient composition

Type of food	Reduced components	Added components	Flavour effects
Food for diabetics	sucrose dextrose lipids	sorbitol xylitol fructose aspartame saccharin	flavour decrease flavour changes body- or mouth-feeling reduced or lacking off-flavour in some cases
Fat-reduced foods, low-caloric foods	lipids	water emulsifiers	flavour decrease in some cases off-flavour in some cases
Protein-enriched or protein-modified foods		proteins traditional or new types	flavour decrease off-flavour in some cases
High-fiber foods		nondigestible additives: modified natural and other polymers; bran	flavour changes off-flavour in some cases
Sodium-reduced foods	sodium chloride	potassium salts no additive	salty taste reduced or lacking flavour (aroma) sensation decreased off-flavour (potassium salts)

the traditional nutrient composition in normal foods often causes a typical flavour decrease or off-flavour formation. Well-known examples are the replacement of sugar by other sweetening compounds, fat reduction in biscuits, sausages or cheeses, and salt reduction in bread or sausages. In all these cases the flavour change is great enough to prevent a large number of patients from following a proper diet. This unfavourable effect is a consequence solely of unacceptable flavour quality.

Strange flavours transmitted from the raw materials represent another problem connected with the introduction of new food sources or of isolated nutrients. No matter whether they originate from the raw material, the isolation process or the changed nutritional composition, they have the character of 'off-flavours' (see chapter **1.3.2.**). In most cases it is difficult to remove or to mask such types of food damage (H. A. GREMLI 1974, J. SOLMS 1974).

The quality of the flavouring is frequently the most important prerequirement for the acceptability and introduction of new foods. The term 'tailor-made' may often be valid for the physical characteristics of such products. But foods with a new, simulated nutrient composition have no chance on the market unless their flavour quality is adapted to the quality of the traditional product.

Today, however, it is not difficult to find examples of a successful solution to this problem. The replacement of animal protein by vegetable protein, for example in certain types of soups, snacks, or dressings, can result in products of high acceptability. The same is true for dietetic foods intended for diabetics if the sugar is replaced by fructose or xylitol instead of sorbitol, cyclamate or saccharine.

It is also probably true that the consumer can adapt to changed flavour qualities, too. Usually, however, such a process takes time, and it will not always be successful — especially in cases where a new flavour quality is concerned.

On the other hand, we should remember some well-known examples of the rather rapid introduction of new foods. Because of their interesting flavour, cola drinks, passion fruit or snacks based on corn or potato have spread all over the world in a very short time. Some further examples may also support this opinion:

Several years after World War II, unsalted butter was introduced generally in the Netherlands because of better shelf-life. Now people prefer the salt-free product (F. D. TOLLENAAR 1975).

In Central Europe in recent decades rye bread has been substituted by wheat-rye blend bread because of simplified processing in industrial plants. Now a large part of the population in question prefers the new aroma (M. ROTHE 1975).

At present people in Southern European countries consume refined neutral-tasting vegetable oils instead of the previously used raw oils with a special aroma. The refined oils, however, were introduced by industry because of their better storage properties (F. D. TOLLENAAR 1975).

1.5.2. Problems and perspectives in food flavouring

What about the main future tasks of aroma research? The answer must be based on two points. First, in highly developed countries we observe such specialization in all kinds of foods as has never been known before in history. This development continues, as a consequence of the economic, technological and nutritional aspects mentioned above. Flavour problems are a part of it.

Second, we should consider that in modern conditions flavour and aroma problems can be solved in the easiest and most economic way by the use of aroma and flavour concentrates. Although we are just at the beginning of a new area of food production and processing, the use of aroma concentrates and other flavourings is becoming more and more widespread. In Central Europe the sale of such products 20 years ago did not exceed 0.1% of the whole food market (H. U. DAENIKER 1975). At the same time aroma concentrates and taste enhancers reached 15% of the weight and 35% of the value of all food additives in the USA (J. W. ANGELINE and G. F. LEONARDOS 1973). In the early 1970s flavour industry offered 10000 new compositions to the food industry yearly (E. J. EISERLE and W. J. DOWNEY 1971). The market for these products doubles every 10 years.

To begin with, activities in the broadened field of aroma research were concentrated on the qualitative analysis of the aroma spectrum. The reason is clear: anyone who tries to enhance or to substitute traditional flavour has first to know its composition. But the complete elucidation of the aroma complex of single foods is much more difficult than was previously supposed. In most cases several hundred aroma components contribute to the flavour, or are considered to participate (see also table 18 on p. 59).

The idea of producing aroma concentrates made up of single components, however, requires additional quantitative information. In the case of most flavourings, the quantitative ratio of the participating substances decides the acceptance and popularity. Quantitative data are still rare in the literature though are are first summaries (H. MAARSE 1982, 1982, 1985; L. NYKÄNEN and H. SOUMALAINEN 1983). One of the main problems inhibiting the availability of more quantitative data is the fact that before the separation of the aroma spectrum by gas chromatography, time-consuming isolation and concentration steps are often included in the analysis. The separation of water or organic solvents from aroma components, however, is associated with loss of varying amounts due to the differing volatility of the single aroma compounds. Thus chromatographic analyses often deliver comparable, but not quantitative results.

The consequences for the development of aroma concentrates are clear: a successfully realized complete recombination of food aroma complexes out of single components is so far unknown. Nevertheless, modern flavour industries produce many pure aroma components. They can be isolated from special aroma complexes, for example, essential oils, or synthesized chemically from other raw materials (see figure 16). The main application for single aroma components is the improvement of naturally occurring aroma complexes, especially as to their aroma intensity. Key substances of the aroma are of special interest (see figure 39 on p. 60).

The practical method of aroma composition within the flavour industry consists of a systematic sensory testing of combined flavourings and single aroma substances. As a prerequisite for successful work, a 'bank' of aroma substances must be available for the 'flavourist' (a highly sensitive, long-trained

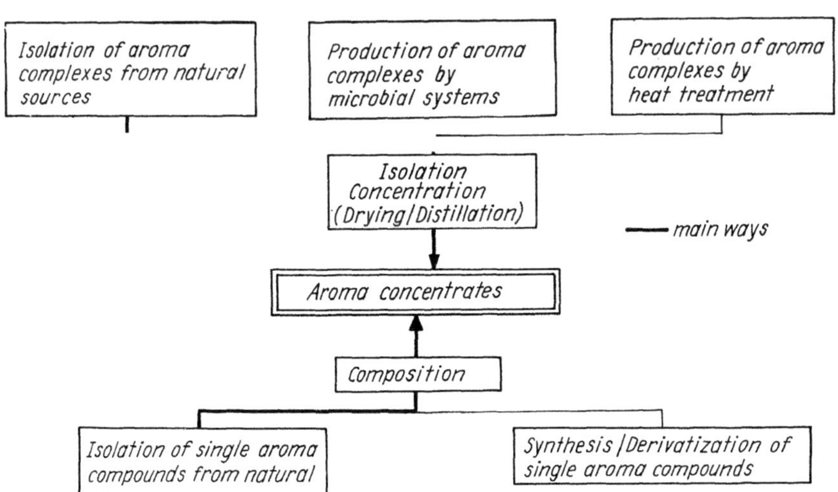

Fig. 16
Main possibilities for production of aroma concentrates

and experienced specialist). For creative work it is necessary to have a good 'aroma memory' (J. MERORY 1968).

That this method of flavour development can be applied successfully is indicated not only by the rich palette of aroma concentrates offered by the flavour industry. There are some preliminary examples of a real simulation of natural aromas which do not belong to the traditional field of flavouring. Thus the imitation of caviar aroma seems to be solved in the USSR (R. V. GOLOVNJA and I. ZURAVLEVA 1970), where flavourized products based on milk proteins are on the market. Finnish research groups are dealing with the imitation of whisky aroma (P. SALO et al. 1972, P. SALO 1973, 1975).

In the past the production of single aroma components by the flavour industry was preferably based on natural material. However, this is economical only if the sources contain high amounts of the components in question. Thus, essential oils, citrus fruits or spices have frequently been used as sources. Several important aroma components are now synthesized chemically by the flavour industry. With regard to food laws identity with the natural product and a high purification level are the main points to be taken into account.

But aroma concentrates must meet high requirements, not only from the qualitative point of view. They must also be highly productive, in order to keep the application level low. Thus, technical concentrations processes are often necessary. As well as traditional concentration methods such as pressing, extraction or distillation, modern drying processes are of great importance. With respect to their aroma intensity, steam distillates (essential oils) are superior to pressed oils or alcoholic extracts. Among the solvents, low boiling ones are preferred on account of their easier separation from the dissolved volatile aroma components. At present, fluorized hydrocarbons (Freons) and carbon dioxide are considered to be especially suitable solvents. Sometimes the pressing procedures formerly used did not greatly influence the natural aroma, which resulted in a better aroma quality.

Aroma concentrates consisting of a restricted number of single compounds frequently develop a less pronounced one-sided quality. A few of them, however, have a high aroma

Table 10
Essential oils important for the food industry and their main components

	Origin	Aroma effective main components	% in essential oil
Anisseed oil	fruits of *Pimpinella anisum*	anethol	80—90
Bitter-almond oil	kernels of *Prunus amygdalus*	benzaldehyde	85
Eucalyptus oil	leaves of *Eucalyptus* sorts	citral, piperitone cineol, citronellal	sort dependant
Hop oil	flower catkins of *Humulus lupulus*	myrcene	30—50
Caraway oil	fruits of *Carum carvi*	carvone	50—60
Cloves oil	flower-bud of *Eugenia caryophyllata*	eugenol	70—90
Peppermint oil	leaves of *Mentha piperita*	menthol	50
Mustard oil	ground kernels of *Brassica nigra*	allyl isothiocyanate	90
Cinnamon oil	bark of *Cinnemon* sorts	cinnemon aldehyde	70—90
Lemon oil	peel of *Citrus limon*	citral	2—6

effectivity. The isolation of single compounds from natural sources has been used by the flavour industry for a long time, for instance in the case of menthol from mint, citral from citrus peel, benzaldehyde from bitter almonds or eugenol from cloves. Today many of these aroma components can be produced more economically by synthesis. In several countries, however, food laws restrict this possibility.

The lower production costs of chemically synthesized aroma components are not their only advantage. Often, synthesized products have a higher degree of purity. Even a partial synthesis of aroma compounds starting from raw materials of natural origin may be economically beneficial. If the food law situation does not prohibit this, the industry will use such processes. We should expect that in future the synthesis of nature-identical aroma components will become more and more important.

The main disadvantage of 'imitation flavours' derived from a limited number of single aroma components is, as already mentioned, the uniformity of the aroma. This can be partially overcome by combination with natural aroma complexes, which deliver the lacking 'background flavour'. In most cases from 5 up to 25% of the total aroma is replaced by the natural aroma complex.

When we consider the high complexity of food aromas, at first sight the isolation of complete aroma complexes from naturally occurring aroma-intensive foods seems to be more profitable. As a principle advantage, such natural products have a high and well-balanced flavour quality. There are, however, some limitations. Most highly developed countries are situated outside the tropical and subtropical zone, and have to import the raw materials. The availability of the aroma materials may depend on such factors as harvest yield, price on the world market, or other economic considerations. Thus there have recently been efforts to develop aroma complexes by industrial processing, using microbial or enzymatic activities or thermic procedures (see figure 16). Cheese and butter aroma concentrates produced in a controlled fermentation system, as well as meat-like flavourings produced by means of a controlled non-enzymatic browning reaction, are early examples. Basic research has also begun in the field of plant cell cultures and their future application for aroma production (H. Ruttloff and M. Rothe 1985; R. J. Whitaker et al. 1986).

In general the natural aroma materials mentioned, as well as the aroma complexes produced by enzymatic or thermic technologies, are only of low flavour intensity. In order to increase their effectiveness, concentration procedures are required. In most cases water has to be removed, with the minimum possible loss in volatile aroma. Aroma quality and cost factors play a dominating role. Lyophilization, for instance, is highly appropriate for aroma retention, but it is also the most expensive process. Thus, in most countries lyophilization of foods is restricted to expensive products like coffee, some fruits or spices. As for other drying techniques, spray drying in the presence of carrier substances is a valuable process frequently used for the concentration of aroma-rich materials.

Among the various modern concentration processes for aroma concentrates, the distillation principle developed in 1944 in the USA for fruit juices (H. P. Milleville and R. K. Eskew 1944, 1946) is still dominant. In this process the high volatility of the most aromatic components is used to separate them with the first 10—40% of the water-rich distillate, and if necessary to concentrate it (see also J. D. Johnson and J. D. Vora 1983). After separation of the aroma, highly efficient concentration steps can be applied to allow the main part of the material to be dried. After separate storage recombination of the aroma and concentrated juice fractions results in a product of high aroma quality (F. Emch 1967). Fruit aroma concentrates of this type have been successfully used for flavouring purposes.

As already mentioned, lyophilization of foods rich in flavour attributes is not only the most favourable concentration process, but also the most expensive one (see figure 17). The reason for a relative high aroma retention in this process despite the simultaneous volatilization of high amounts of water is the object of much discussion. Nevertheless, it is obvious that there are interactions between the volatilized aroma components which are condensed in the form of droplets on the surface of solids surrounding them (mechanical inclusion in micro spaces: J. M. FLINK 1972, M. KAREL 1973; J. SZEJTLI; molecular diffusion: C. J. KING and H. A. MASSALDI 1974, P. J. A. M. KERKHOF and H. A. C. THIJSSEN 1975).

Reverse osmosis and freeze concentration (separating water by freezing: see F. DRAWERT et al. 1981) represent new developments in the field of drying processes with high aroma retention. But here also costs are high. The preferred type of drying process depends on these costs as well as on the aroma loss or change. J. L. BOMBEN et al. (1973) reviewed research work in this field in detail.

The above-mentioned use of microbial or thermic systems for the production of whole aroma complexes is a recent field of flavour production, in which activity is increasing. Meat-like flavourings are produced by a controlled MAILLARD reaction working in the presence of sulphur compounds like cysteine (L. SCHUTTE 1974, R. A. WILSON and I. KATZ 1974, 1980; R. SCHRÖDTER et al. 1986).

Special types of chicken flavouring can be produced in a similar way using partially oxidized fats as reactive compounds (J. POKORNY et al. 1973, 1980; R. SCHRÖDTER et al. 1986).

Early examples of the practical use of enzymatic/microbial aroma formation are to be found in the milk processing industry. The production of Roquefort cheese aroma concentrates by biotechnological fermentation of a milk medium with *Penicillium roqueforti* may be mentioned here. The reaction conditions are focussed to a rapid aroma formation using controlled liquid systems (B. K. DWIVEDI and J. E. KINSELLA 1974, J. H. NELSON 1970, M. ROTHE et al. 1984). Moreover, the addition of microbial lipases and esterases to milk products in order to improve and intensify flavour quality is an often discussed point in recent literature (E. W. SEITZ 1964, R. G. ARNOLD et al. 1975, I. L. GATFIELD 1986, K.-C. M. LEE et al. 1986).

Biotechnological aroma production, however, may provide many examples in future. Fer-

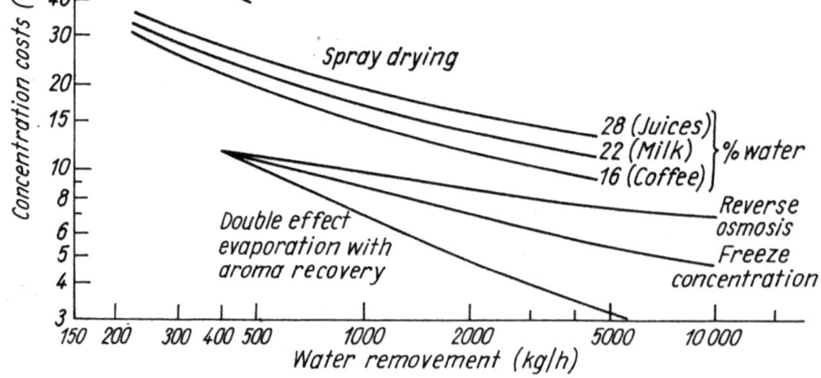

Fig. 17
Cost of concentration and drying for 7500 hours per year in dependance on the drying procedure
(J. L. BOMBEN et al. 1973)

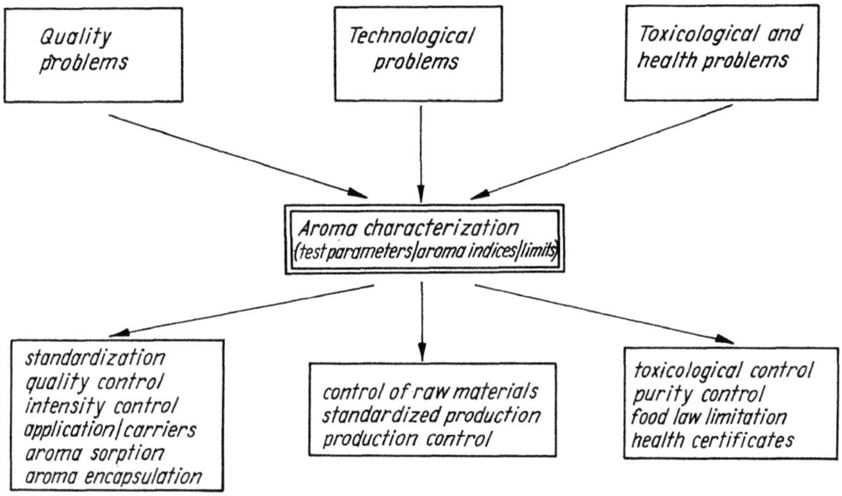

Fig. 18
Frequent problems within production and use of aroma concentrates

mentations flavours play an important role in the production of numerous foods, such as alcoholic beverages, bread, cocoa and butter. If modern production methods reduce the chance of developing the traditional flavour quality within these products, the use of aroma concentrates could represent a possible way of changing this situation (H. Ruttloff 1983).

Apart from the traditional area of flavouring (sweets, confectionery, beverages, margarine) the application of new aroma concentrates still meets a barrier of additional problems (see figure 18). In the near future these may tend to increase, and therefore they should be taken into consideration as soon as possible in future research work.

The recombination of an elucidated aroma complex, for example, does not fail only on account of the complexity of the aroma spectrum itself. Aroma perception is also influenced by the amount and character of the nutrients which are the main constituents of food. Within solid foods aroma components are not free and available for perception, but are often adsorbed to the solid polymers. In this form volatility and vapour pressure may be limited, and this must reduce the intensity registered from the stimulus.

This aroma-binding effect differs in the degree, and is dependent on the properties of the nutrients which act as aroma carriers. When the effect is not considered, however, the reconstitution of an aroma cannot be successful. Examples of such negative effects in aroma composition work can be found in the literature. D. F. Anderson and E. A. Day (1966), for example, tried to recombine the aroma of Blue cheese, which is relatively uncomplicated. But, although quantitative data were available, a very one-sided aroma resulted, which was far away from the natural one. E. J. Mulders in 1973 tried to achieve a composed wheaten bread aroma by combination of the previously identified aroma components. He carried out the combination in such a way that the concentration in the head space over the medium corresponded to that over wheaten bread, but only a dough-like aroma resulted. This may be caused by unknown binding effects, as well as by the possibly incomplete elucidation of the aroma complex.

The sorption of aroma components on carrier systems is one of the most important problems for the flavouring of newly introduced food constituents. There have been in recent years several interesting publications in this field of interaction dealing with model systems (H. G. Maier 1972, 1974, 1975), solutions (R. M. Pangborn and A. S. Szczesniak 1974, D. G. Land and J. Reynolds 1981), carbohydrates (F. Osman-Ismail and J. Solms 1973, R. Wyler and J. Solms 1981), protein (H. A.

GREMLI 1974, M. BEYELER and J. SOLMS 1974) and purines (B. M. KING and J. SOLMS 1981).

More knowledge in this field is required, not only for the composition work of the flavour industry but also in order to solve problems of application, as well as of storing aroma concentrates. The effectivity of flavourings can easily change in the course of solution or encapsulation processes. On one hand the application of microcapsules as aroma carriers (in practice often solved by variants of spray drying) facilitates transport and application because of the solid and free-flowing consistency. On the other hand, the aroma included in the capsules is protected against oxidation and volatilization as normally less than 10% of the enclosed material remains on the surface of the carrier (B. A. GUBLER et al. 1974, W. M. MCKERNAN 1972). The encapsulation of aroma concentrates presupposes that the capsule material is easily dissolved during food preparation or the chewing process, in order to liberate the aroma completely (H. G. PEER and B. HOOGSTAD 1975).

Another kind of aroma protection is the use of package materials which prevent aroma permeation. Figure 19 demonstrates some differences between various foils used as food packaging materials. Correct selection of these materials is obviously important for aroma retention.

In today's modern civilization, one of the most frequently and critically discussed problems is the protection of the consumer against toxic substances from the environment. The flavouring of foods is included in this discussion because it deals with food additives. Thus there is a lot of food legislation, and many toxicological considerations may put obstacles in the way of new foods with a changed composition.

In recent years some advances have been made towards standardization of food legislation in the world. So far, however, there are still differences in the national situation in various countries, and only some general aspects can be discussed here. According to the definition of flavourings and aroma concentrates we can differentiate between natural, nature-identical and artificial flavourings. Nature-identical ones are normally produced by chemical synthesis, but are absolutely identical with naturally occurring substances. By artificial flavourings we mean all those which up to now have not been found in nature.

Within the different codes of flavour legislation, three principles of control in this field can be found:

- In the **restrictive list** system all flavouring additives can be used which are not explicitely prohibited. These prohibited substances are listed.
- In the **positive list** system only those flavouring additives can be used which are included in a positive list. No further substances are permitted.
- The **mixed list** system combines both types, by means of permission and restriction in different ways.

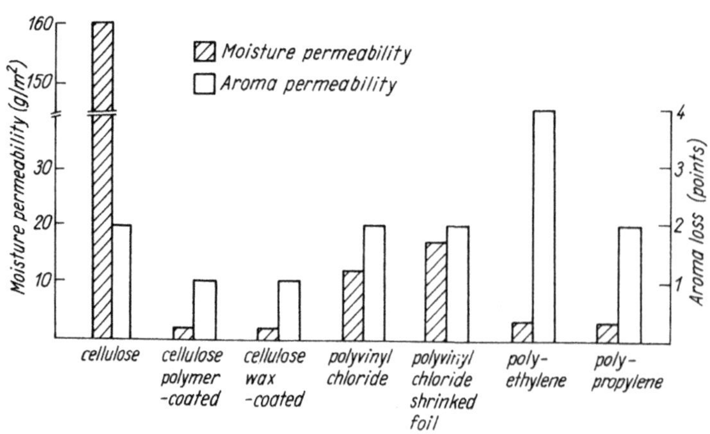

Fig 19
Aroma permeability and moisture permeability of different foils used for food package (E. SCHULTZE 1968)

In the United States the 'positive list' system was introduced many years ago. These lists were established by an independent commission of the American Flavouring Extract Manufacturers' Association (FEMA). They are supplemented and improved continuously. For all flavourings listed there is a given maximum dosage for the different application purposes. Within this area they are designated as GRAS (generally recognized as safe).

The disadvantage of such a system becomes obvious if we consider that at present more than 1500 aroma compounds are included in these lists (R. L. HALL and B. L. OSER 1965 and supplements). The real number of aroma components occurring in foods was said to be between 5000 and 10000 (F. RIJKENS and H. BOELENS 1975). For practical reasons a toxicological testing of such a high number of substances is impossible. This situation differs from that of other food additives. By 1975 the number of all aroma components permitted in the United States was ten times higher than that of all other food additives together (T. COOMES 1975).

The 'restrictive list' system is normally used in such a way that all nature-identical aroma components are generally permitted for application. This concept, which is in principle included in the food laws of several European countries, seems alone to be insufficient, too. The following arguments may be put forward against this concept:

- Nature-identity does not mean that all substances occurring in food are safe. There are also examples for the presence of poisons in some types of foods. Safety for health especially depends on the concentration added.
- The identification of an aroma compound in any food may allow its use in all other foods.
- The system would freely permit the use of an increasing number of additives (F. PETUELY 1973).

In this situation the 'mixed list' system has the greatest advantages. In this system, natural and nature-identical flavourings can be controlled, for instance by means of 'restricted lists' which include critical substances and which may be expanded if necessary. On the other hand, the use of artificial flavourings — if they cannot be abandoned — will be restricted by 'positive lists'. The 'mixed list' system is supported by the flavour industry, too (T. J. COOMES 1975, I. O. F. I. 1976, R. GRÜNDEL and H. E. MUERMANN 1981).

The legislation of flavouring is closely connected with the question of healthy safety and toxicology. Results of animal tests in this field carried out during the last three decades stress the necessity of controlling additives. So far, however, the effects found have been the subject of adverse criticism, for example in the cases of coumarine (C. A. VODOZ 1969), artificial sweeteners (E. GÖTTINGER et al. 1968, H. G. CLASSEN and P. MARQUARDT 1972) or glutamic acid (J. W. OLNEY and L. G. SHARPE 1970, B. I. OSER et al. 1975).

Table 11
Ranking of toxicologically tested aroma components and essential oils in the toxicity scale

Grade of toxicity	LD_{50} (mg/kg body weight)	Percentage ranking in toxicity scale	
		194 compounds tested by C. A. VODOZ 1967	27 essential oils tested by E. VON SKRAMLIK 1959
Extremely toxic	1	—	—
Strongly toxic	2—50	—	—
Medium toxic	51—500	5%	3%
Lowly toxic	501—5000	57%	37%
Practically non-toxic	5001—15000	30%	60%
no effect	15000	8%	—

The real danger, however, seems to be low or non-existent. On account of their high efficacy, aroma components can normally be used only in very small amounts. For toxicity they have to be tested in very much higher amounts. If the LD 50 is used as a tool for describing the toxicity of flavourings, the results in table 11 indicate a very low value. The data are based on 194 aroma substances tested by C. A. VODOZ (1969) and 27 essential oils tested by E. VON SKRAMLIK (1959).

Though toxicological tests are very time-consuming and involve extremely high costs they are increasingly employed in the safety control of flavourings. D. L. J. OPDYKE published summarized results (1972—1975, 1981). Because of the increasing burden of such problems, international cooperation should be sought in future.

2. Aroma analysis

2.1. Sensory analysis

2.1.1. Sensory testing and evaluation methods

Unfortunately, it is widely believed that a sensory aroma or flavour testing can be carried out at any time, anywhere and by anyone. This erroneous idea originates from the mistaken assumption that everyone, as a consumer of food, ought to be able to make a sensory analysis of food flavours on the basis of life-long experience with foods and meals. But in fact this is only one of several requirements for a successful and correct sensory analysis.

Sensory testing is the most important tool solving aroma problems, as it must always be the first and main measure. The replacement of sensory by instrumental values is not possible without knowledge of the type and significance of correlations between these data.

In order to obtain significant sensory results, many conditions must be met. Though a well-trained test panel may be available, there are still numerous factors which can deviate and disturb the registered sensory data. Only if all the requisite conditions are strictly considered can the formerly used (incorrect) term 'organoleptic test' be thought as a 'sensory analysis'.

Sources for wrong or deviant results originate in the psychophysical area, especially when human subjects are used as an instrument for sensory analysis. This begins with the readiness of the test subjects to identify themselves with the problem, and to deliver good results. Each participant should have enough time for the task of sensory testing. Another important requirement is the absolute exclusion of contacts between the members of the test panel during the session, in order to avoid any influencing of the individual opinion. Many other precautions are known which ought to be considered. Some of them are listed here:

- selection of a favourable time of day
- isolation of the test subjects by cabins or shades
- preparation of the test samples outside the test room or before the test session
- presentation of the test samples in coded vessels and in random order
- exclusion or damping of noise during the session
- opportunities for cleaning the mouth between two sensory tests by neutral food or beverages (soda water, tea, rolls, wheaten bread)
- possibility of reducing illumination, or changing its colour
- possibility of tempering the samples thermostatically (water baths)

possibility of ventilation or air conditioning.

The selection of suitable test subjects and their long-term training represents a further important condition for significant results of sensory analysis. Besides consumer acceptance tests, an examination of the people in question should be obligatory in order to exclude people who deviate in their sensory sensibility (R. NEUMANN et al. 1974, 1983). Suitability for one sensory test cannot be carried over to all other sensory test methods (C. Y. KRAMER 1955, E. H. DAWSON et al. 1963, M. A. AMERINE et al. 1965, G. JELLINEK 1968, 1981, S. L. MARTIN 1973).

Each member of the test panel should be able to fulfil the following criteria (J. WITTES and A. TURK 1968, G. JELLINEK 1981):

- to be motivated for the task in hand, and to have no aversion to the tested product,
- to differentiate between different aroma or taste effects even in a low intensity area,
- to be able to recognize and to define them,
- to single out different attributes in complicated systems,
- to understand the type and requirements of the test method,
- to adapt to the conditions during the session.

In table 12 the rank order of a sensory evaluation of food flavour is summarized. Tasting in the mouth or smelling always starts with the question of whether or not there is an odour or taste, respectively. Subsequently a qualitative analysis is carried on in the brain by comparing the new impression with earlier stored ones. After recognition of the type of aroma, a quantitative estimation and/or hedonic evaluation follows. Finally, the registered impression — whether known or new — is 'computerized' and stored in the memory.

The selection of panel members for a special type of sensory test, and their training for this task, can take different lengths of time depending on the sensory test method involved. Normally it is a matter of weeks. Specialized sensory profile analysis will sometimes have no significant results without several months training (S. L. MARTIN 1973). J. J. POWERS and M. C. QUINLAN (1974) found occassional strong deviation from the average value even with well-trained test subjects. This leads to the conclusion that one should apply more special sensory test methods for the purpose of selecting a test panel. A high sensitivity to all sensory attributes is obviously an exceptionally rare human attribute.

For the selection of panel members, the following criteria should be taken into consideration:

- previous knowledge and training of the test subject (layman, trained layman, panelist, flavourist),
- age and experience,
- sex,
- consumption of stimulants or drugs,
- time of day, saturation, last meal,
- adaption, fatigue state,
- stress, ambitions, aversion.

G. JELLINEK (1981) discusses the necessity of educating sensory panelists, and proposes minimum demonds at the end of the schooling and training period. These are summarized in table 13.

Table 12
Rank order of sensory impressions in the course of sensory evaluation

Step no	Question	Sensory impression and memory input or output
1	Whether?	Reception / Registration
2	What approximately?	Sorting / Sieving
3	What exactly?	Recognizing / Ordering
4	How much?	Comparing / Quantitative estimation
5	How likely?	Evaluating / Hedonic response
6	(For later impressions)	Computerizing / Storing / Input

Table 13
Minimum demand for absolvents of courses for sensory test methods
(G. JELLINEK 1981)

Tasks	Minimum right answers
Recognition of the four basic tastes	100 %
Threshold recognition	80 %
Odour recognition tests	80 %
Aroma recognition tests	70 %
Paired difference tests with special question	70 %
Triangle tests	70 %
Ranking test concerning 4 basic tastes	60 %
Colour ranking tests	90 %
Tactile ranking tests	80 %
Dilution profile tests with aroma	recognition

Within the last three decades there has been a great increase in the literature on sensory analysis. B DRAKE and B. JOHANSSON (1969) registered 2200 publications between 1950 and 1968, and summarized synopses of the contents. F. KIERMEIER and U HAEVECKER (1972) discussed 1135 publications connected with sensory evaluation of foods. According to D. J. TILGNER (1974), between 1958 and 1970 14 international symposia about sensory problems were held.

From this large field an introductory work like this can only discuss some of the most important aspects, and this chapter therefore has to be limited to the principles of the sensory methods (the reader may find further details in the publications of J. F CAUL 1957, D. J. TILGNER 1962, R. M. PANGBORN and S. C. OUGH 1965, G. JELLINEK 1964, 1968, 1981, 1984, L. KIERMEIER and U. HAEVECKER 1972, H. MOSKOWITZ 1974, 1975, 1978, R. NEUMANN et al. 1974, R. NEUMANN et al. 1983).

In table 14 the most commonly used sensory test procedures are summarized in a schematically shortened form (for details see M. A. AMERINE et al. 1965, F. KIERMEIER and U. HAEVECKER 1972, G. JELLINEK 1981, IFT Sensory Evaluation Division 1981, R. NEUMANN et al. 1983). The selection of an appropriate method depends on the problem to be solved. The question to be answered determines the number of samples, of test subjects and of repeated tests.

Because of the high sensitivity of human sense organs to differences in aroma, taste or consistency so-called **difference tests** are widespread in their application to aroma research and quality evaluation of foods. Requirements for the panelists and preparation time are relatively low. Because of these advantages, 'pair-test', 'duo-trio-test' and 'triangle test' are frequently used. Statistical analysis of the results is necessary. Separate judgements must be independent (R. HARPER 1982).

Ranking tests are used especially in such cases where more than two samples have to be compared. The order can be of interest for the intensity of special sensory attributes, as well as for acceptance and consumer preference. Ranking tests are appropriate for the selection of variants of special foods, for optimization procedures related to flavour or for the selection of panel members and panel training.

Threshold determinations can be carried out as difference tests or ranking tests. They are used mainly for determining the efficacy of aroma and taste components (D. G. GUADAGNI et al. 1972, M. ROTHE et al. 1972, F. A. FAZZALARI 1978) or for determining the sensitivity of individual test subjects.

Scaling tests (scale and categories) represent a commonly applied procedure for food quality evaluation. The British Methodology Document of 1980 defines 'classification', 'rating', 'scoring' and 'grading'. In most countries the quality control of foods and beverages by factory or state control groups is based upon this type of sensory testing. The number of points in the scale differs in various countries, although in most cases 5-, 7- or 10-points are used for determination of flavour quality and its different attributes. Appropriate attributes, as well as their weighting within total evaluation values are chosen on the basis of long-term expert experiences.

The weighting, however, is normally an open problem. Statistical methods can be used as a tool (L. TELEGDY-KOVATS et al. 1970, F. ÖRSI 1973, R. NEUMANN et al. 1983). Because of high weighting factors, aroma

Table 14
Most important test methods of sensory analysis and application field
(F. Kiermeier and U. Haevecker 1972; R. Schrödter 1976)

Method	Type and number of panel members[1]	Test sample arrangement[2]	Type of statistical evaluation	Application field
Single sample testing	u: 80 t : 3—10	A, B ...	Analysis of variance	Pre-testing Quality information Preference Degree of liking
Paired test	u: 80 t : 8—12	KA AB AK BA AK AB	Binomial distribution Chi square test Sequential analysis	Difference testing Sensitivity test Panel selection Threshold determination
Duo-trio test	t : 8—10	AK, KA, KA → K	Binomial distribution Chi square test Sequential analysis	Difference testing Sensitivity test Panel selection
Triangle test	t : 8—10	AAB BBA ABA BAB BAA ABB	Binomial distribution Chi square test Sequential analysis	Difference testing Sensitivity test Threshold determination Panel selection
Tetrade test	t : 8—10	KAB, KKA, BAK → K	Binomial distribution Chi square test Sequential analysis	Difierence testing Sensitivity test Panel selection
Ranking test	u: 80 t : 3—10	EGACFBD ↓ ABCDEFG	Rank analysis Analysis of variance t-test	Sample selection Quality comparison Acceptance test Preference
Multiple comparison test	t : 5—10		Analysis of variance Rank analysis	Difference testing Product development Quality control
Rating test (category or ratio scaling, magnitude estimation)	t : 5—10	A, B ...	Analysis of variance Rank analysis	Quality evaluation Quality control Preference
Flavour profile method	t : 6—12	A, B ...	Analysis of variance Rank analysis Graphic presentation	Detailed quality determination Product development Quality control Aroma index
Sensivity test	u: 5—15	FBADEC ↓ ABCDEF	Sequential analysis	Threshold value Panel selection Aroma index
Acceptance test (Consumer test)	u: 25—100	A, B ...	Analysis of variance Rank analysis	Consumer testing Product selection Preference Degree of liking Hedonic response

[1] u = untrained persons; t = trained persons
[2] A, B, C ... = test samples; K = control, standard

evaluation is predominant within total flavour evaluation. In most foods and beverages the sum of (weighted) aroma and taste points exceeds 50% of the total value.

On account of the application of ratio scales **magnitude estimation** has in principle to be included in the group of scaling procedures. Originally developed for the evaluation and calculation of loudness, this method has attracted increasing interest in the aroma field (H. R. MOSKOWITZ 1974, 1975, 1978, R. HARPER 1982, P. JOUNELA-ERIKSSON 1982). The main difference from normal scales is the use of magnitudes, with each one representing one point on the scale. The test panel is asked to define a difference in aroma or flavour intensity by a multiplication factor compared with the standard, which is fixed as a value of 1. Thus, the next point on the scale is the magnitude 10: this means that the sample tested is 10 times as intensive as the standard.

Magnitude estimation is based on the Power Law of S. S. STEVENS (1956) which defines a double logarithmic relationship between concentration and stimulus. The function, however, seems only to approximate the real situation (K. HOPPE 1980). The exponent in the function ($I = K \cdot c^n$) varies within a rather large range.

Sensory profile analysis first introduced in aroma research by S. E. CAIRNCROSS and L. B. SJÖSTRÖM (1950), is a special type of descriptive analysis which not only quantifies total flavour impression but also comprises the intensity testing of several attributes (sensory notes). This type of analysis has a very high predictive value, but demands the highest grade of panel selection and panel training. One especially useful possibility of profile analysis is the graphical demonstration of the results. The graphical profile, in the form of a half or full circle, shows the complexity of most food flavours as well as the composition of the overall impression by single sensory attributes. Such graphical profiles may be thought of as being like a mirror which transforms and transmits a complex shape to our memory.

The graphical profiles first discussed in detail by J. F. CAUL (1957) and D. J. TILGNER (1958) had the character of a half-circle profilogram, and included the ordering of the registered sensations into the analysis and the graphical picture (see figure 20). The order of reception can be used as an additional tool for describing the overall impression. It may be of particular interest in cases where pre- or after-taste is observed. Generally, however, this examination needs a well-trained panel and is not easy to realize.

Today, intensity decriptions of the different notes dominate in literature dealing with this topic. Circular profilograms were first published by H. STONE et al. (1974) and J. M. MECREDY et al. (1974). During recent years the number of publications in this field has increased (K. ZOOK and C. WESSMAN 1977, J. POKORNY et al. 1981, M. ROTHE et al. 1981, M. GILETTE 1984). The advantage of sensory profile analysis becomes evident when we compare it with the simple numerical scales of traditional food quality evaluation. Flavour profiles not only indicate changes or differences in flavour quality, but simultaneously show their causes. Moreover, the specified sensory attributes of sensory profiles have been used successfully in the search for instrumental indices which are correlated with flavour, aroma and their characteristics

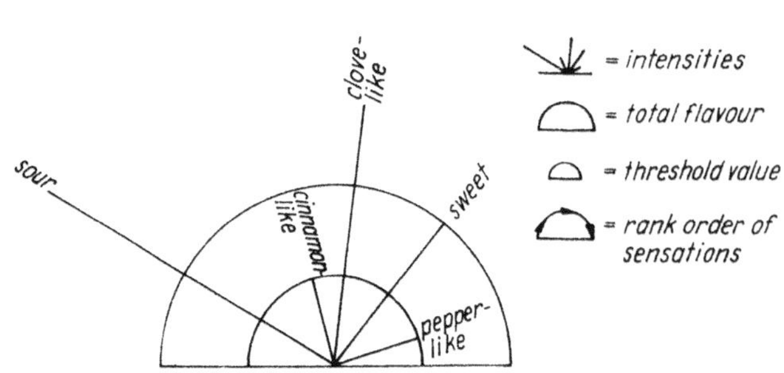

Fig. 20
Aroma profile of a tomato ketchup (D. J. TILGNER 1962)

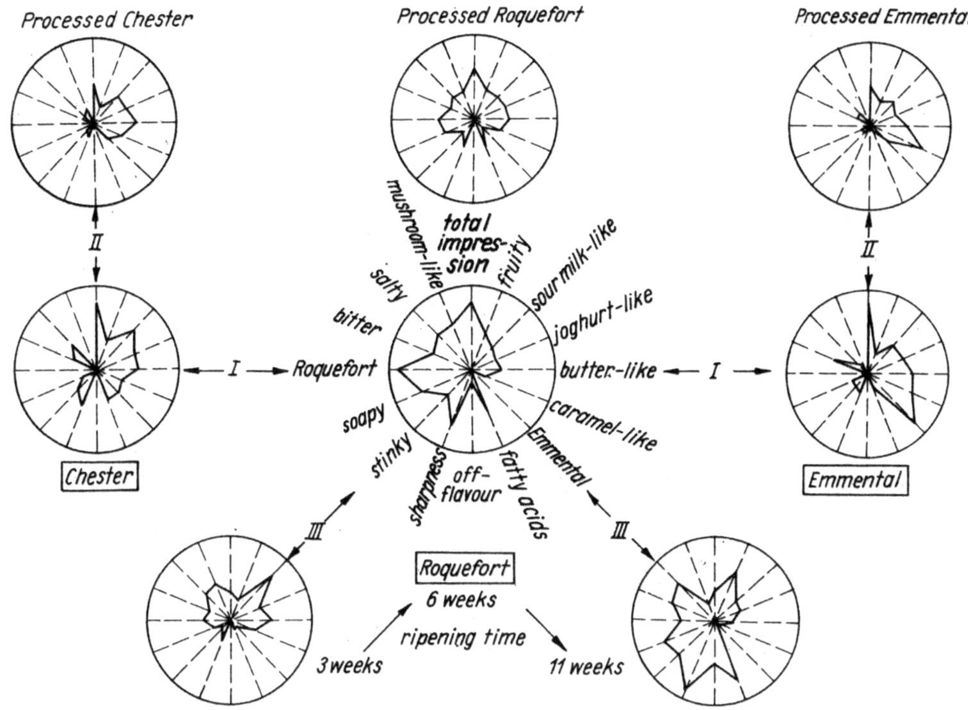

Fig. 21
Characterization of cheese aroma problems by sensory profilograms
(M. ROTHE et al. 1981; I = different sorts, II = processing, III = ripening)

(E. VON SYDOW et al. 1970, T. PERSSON et al. 1974, A. A. WILLIAMS 1975, M. ROTHE et al. 1980).

As an example of the application of sensory profilograms, figure 21 demonstrates differences caused by processing of several cheese types and during ripening of Blue cheese. The changes shown in the upper part indicate the loss in intensity of different types of cheeses during processing. In all three cases the intensity of the sort-typical notes is reduced by the heating process which accelerates the mild caramel-like or butter-like notes only. The ripening study on Blue cheese shown in the bottom part of the figure indicates an optimum quality after five weeks, while before that time the raw, sour milk-like notes, and later on the sharp attributes, dominate.

Selection and training of panels for such profiling procedures are most important for successful research. K. ZOOK and C WESSMAN (1977) use triangle tests for a screening of prospective judges selecting, for example, 10—12 out of 24—36 original applicants. They propose a minimum training of 10 hours, usually 1 hour per day. During this time the panel becomes familiar with the product as well as with the special problem under consideration.

One of the main points of dicussion and training is a good agreement on the understanding of the different sensory attributes. All members of the panel have to characterize the same attribute with the same term. The palette of received impressions, however, can be very broad. Figure 22 demonstrates this fact, using the example of a cider. The graphical picture, taken from A. A. WILLIAMS (1975) indicates relations between different sensory notes and interactions.

For every new problem appropriate sensory descriptors must be found anew, so the success of a profile analysis also depends somewhat on the selection of these terms. As a tool in this work lists of descriptors have been developed. A first such collection of descriptors formerly applied was published by R. HARPER et al. (1968). It was supplemented by

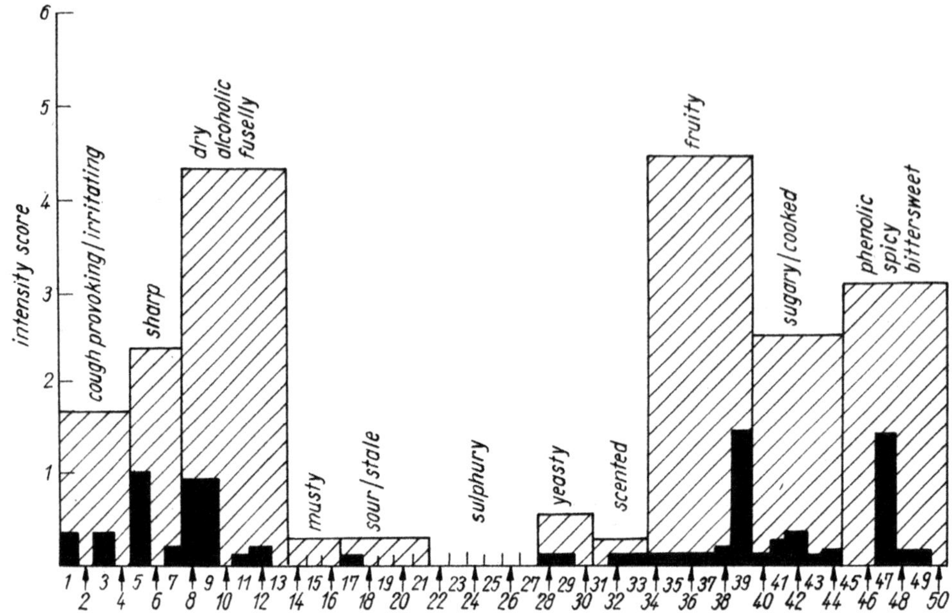

Fig. 22
Profile diagram of a cider
(A. A. Williams 1975 a, b)

1 = like sulphur dioxide
2 = like high concentration of acetaldehyde
3 = like high concentration of amylalcohol
4 = like high concentration of acetic acid
5 = fruit-like sharpness
6 = like rhubarb
7 = like acetic acid (sharp character)
8 = like ethanol
9 = like higher alcohols
10 = like ethyl acetate
11 = like low concentration of acetaldehyde
12 = green (fruity)
13 = green (vegetable-like)
14 = earthy
15 = cardboard
16 = mousy
17 = like acetic acid (sour character)
18 = rancid
19 = old horse
20 = fatty
21 = soapy
22 = rubbery
23 = like hydrogen sulphide
24 = like cooked cabbage
25 = catty
26 = like onions
27 = like shrimps
28 = like mushrooms
29 = like yeast extract
30 = like dough
31 = like cucumber
32 = like roses
33 = floral
34 = like strawberries
35 = like bananas
36 = like pineapple
37 = like pears
38 = like pear-drops
39 = like apples
40 = like cooked fruit
41 = like raisins
42 = like diacetyl
43 = like caramel
44 = like black treacle
45 = phenolic/carbolic
46 = pharmaceutical
47 = like bittersweet cider
48 = spicy (1)
49 = spicy (2)
50 = like celery

A. Dravnieks (1975) and now contains 136 terms. This list discussed by L. J. von Gemert (1981), too, is reproduced in table 15.

Another way of facilitating the understanding of terms within the sensory group is the use of standard odorants, which must be available during the sensory session. G. T. Zausch (1970/71) and R. V. Golovnja et al. (1985) propose a list of standard substances producing typical odour sensations for comparison as well as for panel selection purposes.

The dilution profile method published by D. J. Tilgner (1962) represents a variant of the profile or descriptive analysis. By use of a continuous dilution with water, sensory attributes can be recognized which are masked

Table 15
List of descriptors for characterization of various types of odours
(A. DRAVNIEKS 1975; L. J. VAN GEMERT 1981)

Fragrant	Oily, fatty	Aromatic	Fruity (citrus)
Sweaty	Like mothballs	Meaty (cooked)	Fruity (others)
Almond-like	Like gasoline, solvent	Sickening	Putrid, foul, decayed
Burnt, smoky.	Cooked vegetables	Musty, earthy, moldy	Woody, resinous
Herbal, green, cut grass	Sweet	Sharp, pungent, acid	Musk-like
Etherish, anesthetic	Fishy	Camphor-like	Soapy
Sour, acid, vinegar	Spicy	Light	Garlic, onion
Like blood, raw meat	Paint-like	Heavy	Animal
Dry, powdery	Rancid	Cool, cooling	Vanilla-like
Like ammonia	Minty, peppermint	Warm	Fecal (like manure)
Disinfectant, carbolic	Sulphidic	Metallic	Floral
Perfumery	Yeasty	Eucalyptus	Strawberry-like
Malty	Cheesy	Buttery	Stale
Cinnamon-like	Honey-like	Like burnt paper	Cork-like
Popcorn	Anise (licorice)	Cologne	Lavender
Incense	Turpentine (pine oil)	Caraway	Cat-urine-like
Melona (cantaloupe, honey-dew)	Fresh green vegetables	Orange (fruit)	Bark-line, birch bark
Tar-like	Medicinal	Household gas	Rose-like
Peanut butter	Celery	Leather-like	Nutty (walnut etc.)
Violets	Burnt candle	Pear (fruit)	Fried fat
	Mushroom-like	Stale tobacco smoke	Wet paper-like
Wet wool, wet dog	Pineapple (fruit)	Raw cucumber	Coffee-like
Chalky	Fresh cigarette smoke	Raw potato	Peach (fruit)
Mouse-like	Laurel leaves	Beery (beer-like)	Oak wood, cognac
Pepper-like	Scorched milk	Cedarwood-like	Grapefruit
Bean-like	Sewer odor	Coconut-like	Grape-juice-like
Banana-like	Sooty	Rope-like	Eggy (fresh eggs)
Burnt rubber	Crushed weeds	Seminal (sperm-like)	Bitter
Geranium leaves	Rubbery (new rubber)	Like cleaning fluid (carbona)	Cadaverous (like dead animal)
Urine-like	Bakery (fresh bread)	Cardboard-like	Raisin-like
Lemon (fruit)	Seasoning (meat)	Crushed grass	Maple (as in syrup)
Dirty linen-like	Apple (fruit)	Chocolate	Hay
Kippery (smoked fish)	Soupy	Molasses	Kerosene
Caramel	Grainy (as grain)	Sauerkraut-like	Clove-like

by other components in high concentration. The order of sensations can also be changed during dilution.

Acceptance tests (consumer tests) are of special importance during the development of new food or beverage products, or in the case of a flavour change of the end products caused by new technologies. Food acceptance, however, is influenced by many personal experiences which can vary widely depending on age, sex, consumption habits, social and nutritional aspects (see also figure 69 on page 96). Thus to provide a significant answer to the question of whether a new product will have a chance on the market or not, a lot of people representing the group to be tested have to give their opinion. Judging recent literature, the minimum is 80 people.

In contrast to all other types of sensory procedures, the people used for acceptance or preference tests must not be modified by training. All effective tests measuring preference, degree of liking or disliking, or pleasantness belong to that group of sensory test methods as well as hedonic procedures which ask for a choice between food products of different quality (G. JELLINEK 1968, IFT Sensory Evaluation Division 1981).

2.1.2. Sensory versus instrumental analysis

It was remarked earlier that sensory analysis has a central place in aroma research. This is not only true for food flavour quality, which is generally evaluated by means of sensory methods. All research problems in the field of aroma and flavour have also to deal with sensory methods and sensory evaluation. Without sensory data none of the modern chromatographic techniques can be successfully used. Knowledge of the sensory relationship is always necessary.

Any discussion which comes to a contrary conclusion about sensory and instrumental research is useless and wrong. There is little point in discussing whether the gas chromatographic detector or the human nose is more sensitive, or whether the human brain or a modern computer is superior for the integration process of sensations produced by a complicated multi-component aroma system. It is the combination of both techniques that will bring about progress.

In relation to sensitivity, modern gas chromatographic techniques with a minimum detection level of 10^{-12} reach the same low level which can be smelled by the human nose in some cases of highly effective aroma components. On the other hand, our sense organs need an unbelievably short time for the complex registration and evaluation of hundreds of parallely acting aroma compounds. Within a fews seconds we are able to register, rank and recognize a sensory impression, to compare it with earlier impressions of the same type and store the new sensation (G. T. ZAUSCH 1971).

One important point should always be considered in sensory analysis. When we compare

Table 16
Detection limits of human nose and gas chromatographic detectors for aroma components (E. VON SYDOW 1975; order of magnitudes in g)

Compound	Human nose	Gas chromatography		
		FID	FID-N	FID-S
n-Decanal	10^{-10}	10^{-10}	—	—
n-Hexanal	10^{-8}	10^{-10}	—	—
Ethyl butyrate	3×10^{-8}	10^{-10}	—	—
Methyl mercaptan	10^{-10}	2×10^{-10}	—	3×10^{-11}
Dimethyl-disulfide	3×10^{-9}	2×10^{-10}	—	3×10^{-11}
2-Methoxy-3-isobutyl-pyrazine	10^{-12}	10^{-10}	2×10^{-11}	—
Methyl anthranilate	5×10^{-10}	10^{-10}	5×10^{-11}	—
Pyridine	10^{-8}	10^{-10}	2×10^{-11}	—

our human senses with an instrument like a gas chromatograph, we must compare them under the same conditions. Thus the evaluation of sensory data has to include statistical methods, just as in the case of instruments: error and repeatability, standard deviation and correlation have to be considered and calculated. In the past it was mainly lack in attention to these points which unjustifiably brought sensory analysis into discredit.

As well as brief information about the test procedure itself, table 14 (p. 33) contains short hints for the most advantageous statistical method to be used. Some of the most important points are the necessary proof of a normal distribution of the data obtained, or tests for the significance of calculated differences by variance or other types of statistical analysis. In the literature, tables with limit values facilitate the evaluation of data (see, for example, M. A. AMERINE et al. 1965, G. JELLINEK 1964, 1981, R. NEUMANN et al. 1983).

Statistical evaluation is a special field which can be only briefly mentioned here. As an example, we may consider sequential analysis which is mentioned often in recent sensory literature. It is used for the selection of panel members, or for threshold determination. Such tests are carried out with a pair of test samples which are characterized by a small sensory difference. The samples are offered to the test subject in a random order. The subject under test has to state only whether there is a difference between the first sample and the second, the second and the third and so on. Based on the number of right or wrong answers, the graphical evaluation indicates after a series of such comparison tests whether the tested person is appropriate for this type of sensory analysis or not (see figure 23). In the same way it can be proved, for example, for the purpose of threshold determination, whether the concentrations under test are significantly different.

A special problem related to sensory and instrumental analysis is the question of whether food quality control, currently carried out mainly by means of sensory testing, may be replaced or supported by instrumental methods. This idea seems to be one of the most widely discussed aims of flavour research, but because of the complexity of food flavours we are still far away from this target. Only if key components occur within the aroma complex does the application of instrumental methods for quality control have real chances.

Figure 24 shows one of the earliest examples of a combined sensory/instrumental flavour

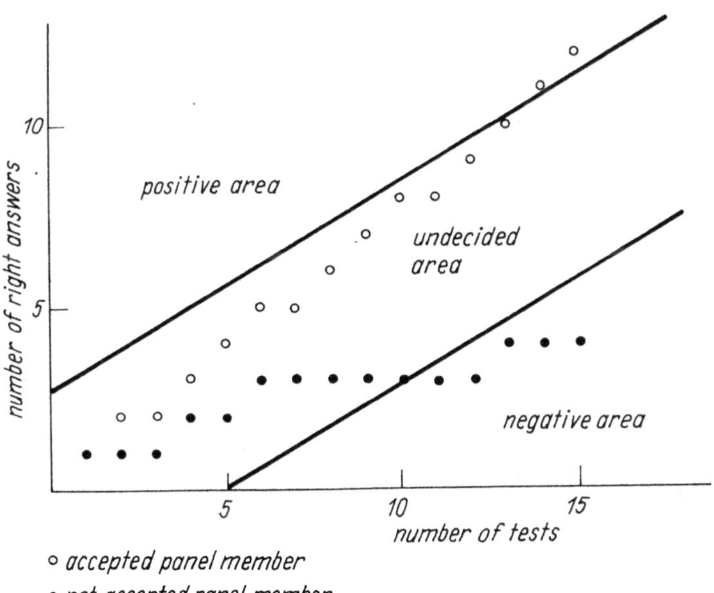

Fig. 23
Selection of panel members by sequential analysis
(M. A. AMERINE et al. 1965; G. JELLINEK 1981)

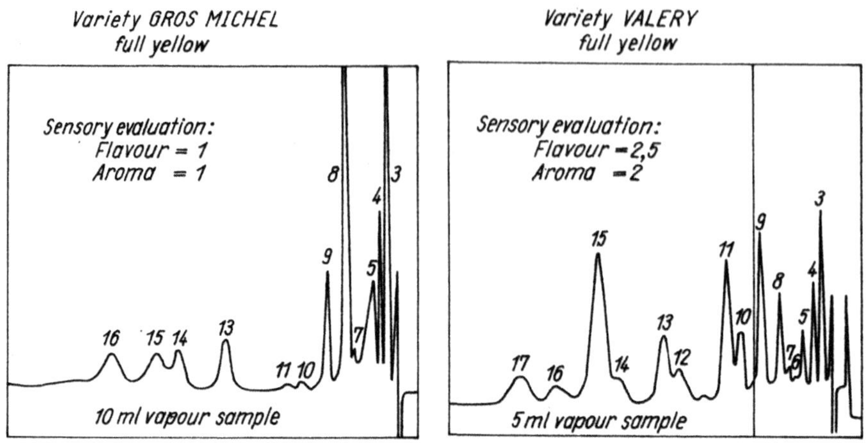

Fig. 24
Combined evaluation of sensory and gas chromatographic data on example of ripe bananas (A. J. McCarthy et al. 1963)

evaluation. A. J. McCarthy et al. (1963) tried to identify typical components within the banana aroma complex by a combined sensory and gas chromatographic technique. Some of the results are shown schematically. As the result of many tests comparing the amount of esters, alcohols and ketones in the head space over bananas in different varieties and ripening stages, it was stated that the typical banana-like aroma is especially related to the content of some amyl esters of short-chain monocarboxylic acids. On the other hand butyl acetate, butyl butyrate, hexyl acetate and amyl butyrate were found to be responsible for the 'fruity', 'ester-like' note.

Since then there have been many more examples of the successful correlation of sensory and instrumental data, indicating the possible success of such combined work. Some cases will be discussed in the following chapters.

2.2. Chromatographic analysis

2.2.1. Development of chromatographic techniques

Before 1950, the number of aroma components identified in single foods was limited to less than 10, with the exception of a few highly flavoured products (coffee, spices, essential oils). A very few number of photometric methods were then available, which were appropriate for identification of trace components within highly complex mixtures by simultaneous high sensitivity and specifity. But because of the precence of many accompanying disturbing substances in much higher concentrations, every aroma analysis in those early days was a very difficult procedure. There was not only the problem of isolating of the aroma complex from the foods without alteration, but also the fact that there were not enough classical analytical tools for the separation of the numerous volatiles.

It was the development of chromatographic techniques which opened the door to the labyrinth of food aroma composition. The term 'chromatography' is presently applied to all separation processes based on the distribution between a stationary and a mobile phase (S. van Straten 1981).

The technique of chromatography started with the classical experiments of the Russian scientist Tswett (1906) who used **column chromatography** for the first time in his work on the separation of natural colourings. However, another 30 years were to pass before an effective application of this chromatographic technique. In 1938 N. A. Izmailov and M. S. Schraiber published results achieved by a variant of **thin layer chromatography** which they called the 'drop chromatography method'. The separation of drug plant extracts was carried out on thin layers distributed on glass plates by dropping an ethanolic extract on one point of the sorbent surface and developing the rings formed by additional drops of pure solvent.

Approximately one decade before the wider application of this technique R. CONSDEN et al. added **paper chromatography** to the arsenal of analytical tools in 1944. The great success of this method in the field of amino acid separation brought paper chromatography to the forefront among chromatographic methods in a few years. It was at that time that the palette of known aroma components was first broadened by chromatography. The idea of acquiring more knowledge about food aromas consequently grew in the years after the malnutrition and hunger following World War II. Several of the early methods of paper chromatographic separation of aroma components were focussed on 2.4-dinitrophenylhydrazones, because their intensive yellow colouring indicated the separated spots visually on the paper.

When the application of paper chromatographic methods was at its height, the thin layer techniques finally achieved a break-through. In contrast to paper chromatography, the new variant shows a much higher separation effect and 10 to 100 times higher sensitivity, and works more rapidly (H. W. H. SCHMIDT 1967). I. G. KIRCHNER et al. (1951) were the first to use glass strips with a sorbent surface for separating terpene derivatives. The application of thin layer chromatography was very much assisted by a number of simple pieces of equipment, for example for coating the plates, invented and proposed by E. STAHL (1956). Today the main advantages are the availability of many different separation phases, and a relatively simple performance which does not need expensive equipment.

The most important step forward in separation and identification of aroma components, however, came with the introduction of **gas chromatography**. Activities in the aroma field started a few years after the classical work of A. T. JAMES and A. J. P. MARTIN (1952). This procedure made it possible to split a very complex system such as a food aroma into single components, in many cases within only one separation step. As aroma components have boiling points no higher than 300 °C, it is understandable from this point of view, too, that gas chromatography should have become the major tool in food aroma research (C. WEURMAN and G. VAN LUNTEREN 1967). Finally, it should be remarked that the separation and splitting up of the aroma complex into single components was facilitated by the relatively simple structure of most aroma components. There were therefore excellent opportunities for technical advances such as mechanization, checking and automation of the gas chromatographic apparatus.

Liquid chromatography (HPLC) is the youngest of the various chromatographic separation procedures. Within the aroma field its application is concentrated particularly on problems where substances with high boiling points, or low vapour pressures, or high molecular weights are of interest. These conditions may be true not so much for aroma components themselves as for aroma precursors, taste substances or flavour potentiators, for example. The separation principle is similar to other chromatographic techniques: the separation is carried out on columns with solvents under high pressure. For detection, ultraviolet, fluorescence, infrared or mass spectrometric detectors are available. The easy collection of fractions represents one of its main advantages. In some cases liquid chromatography is preferable to capillary gas chromatography with respect to the separation and solution effect (L. L.-P. WU and C.-M. WU 1981).

2.2.2. Isolation and concentration procedures

Normally the concentration of aroma components in foods is very low. Only in a few cases does the sum of all odorous volatiles exceed 0,1%, for example in such aroma-rich products as coffee (O. VITZTHUM 1976), distilled alcoholic beverages (L. NYKÄNEN and

H. SUOMALAINEN 1983), spices or fruit aroma concentrates (P. DUERR and U. SCHOBINGER 1980). The single volatiles are usually found in amounts of 0.1 to 0.001 p.p.m., some of them in much lower concentration (J. M. H. BEMELMANS 1979). Figure 25 shows the composition of a coffee extract based on dry matter. This graphically demonstrates the fact that, besides the aroma-active compounds, many other substances are present which can disturb the aroma determination and which must be separated before analysis.

Thus in most cases a gas chromatographic analysis cannot be started without preparation work which is sometimes much more time-consuming than the analysis itself. To begin with, sample selection and — if necessary — sample preparation have to be carried out. This is followed by an isolation of the aroma compounds. The type of isolation method used depends on the physical condition, state and composition of the material under test. Figure 26 summarizes five principle isolation techniques. Of these, distillation methods, which use the high volatility of aroma compounds, are most often applied. In second place are extraction procedures, which are based on favourable distribution coefficients between the solvents used and the aroma within the food system. Different methods are applied for solid foods, aqueous products and fats and oils (J. M. H. BEMELMANS 1979, 1981, S. S. CHANG 1973, M. DANIEWSKI and Z. MIELNICZUK 1974, M. M. LEAHY and G. A. REINECCIUS 1984).

Although distillation methods are the predominant isolation procedures for aroma components they have some disadvantages, too. Higher temperatures, necessary for example in steam distillation, must be used with caution. Destruction of components, as well as formation of artefacts, is possible (C. WEURMAN 1969, J. A. PICKETT et al. 1975, M. A. AZZOUZ et al. 1976). Vacuum distillation or gas trapping methods may therefore be preferred in many cases.

Volatilized aroma components are normally condensed in cooled traps (C. WEURMAN et al. 1970) or adsorbed on porous polymers which have begun to replace the charcoals formerly used (J. SCHAEFER 1981, S. NITZ and E. JÜLICH 1984, H. SUGISAWA et al. 1984). Distillation methods have the advantage of separating completely in the first step of the analysis all non-volatile components which are normally present in much higher amounts than the volatiles.

For liquid and aqueous foods an isolation by solvent extraction is frequently preferred. Solvents with a low boiling point are very suitable (diethyl ether/pentane mixtures, methylene chloride, fluorinated hydrocarbons of short chain length, acetonitrile, carbon dioxide). The removal of the solvent, which is necessary in most cases before chromatographic separation, is carried out by distil-

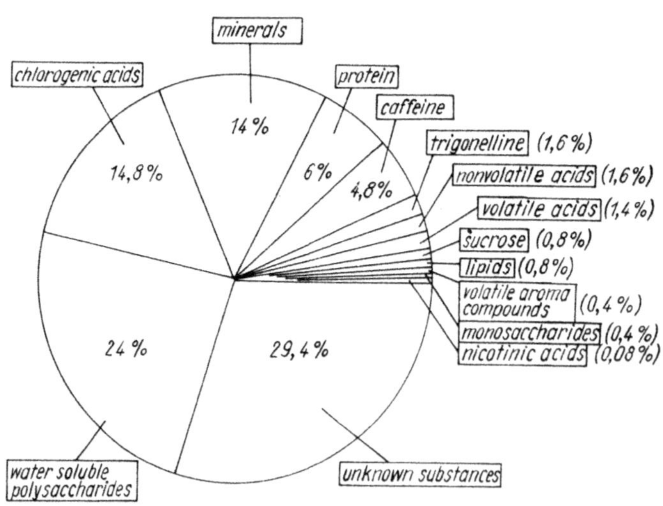

Fig. 25
Composition of a coffee extract (O. G. VITZTHUM 1976)

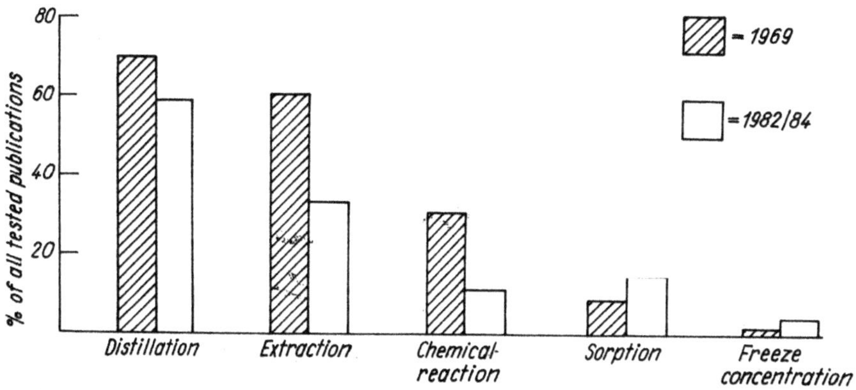

Fig. 26
Application of different isolation and concentrauon procedures for aroma components (1969: evaluation of 234 publications by C. WEURMAN; 1982/84: evaluation of 250 publications selected from Documentation 'Nahrung und Ernährung' of Central Institute for Nutrition Potsdam-Rehbrücke)

lation in a special apparatus. Another possibility is the use of very high-boiling solvents which do not disturb the chromatographic analysis. In this way difficult concentration steps can sometimes be avoided (M. NONAKA 1971).

Normally, concentration follows the isolation step. Solvent, water or the aroma components themselves have to be separated or enriched. Here distillation as well as extraction, chemical formation of derivatives, adsorption or freezing techniques can be used.

For a highly effective selective separation, several methods may be used in combination. In recent years the simultaneous distillation/extraction procedure first discussed by G. B. NICKERSON and S. T. LIKENS (1966) has gained favour with many workers in the aroma field, and has been much used. In the literature several modified versions of this procedure are described (K. WUCHERPFENNIG and G. BRETTHAUER 1976, H. SUOMALAINEN and L. NYKÄNEN 1968, H. MAARSE and R. E. KEPNER 1970, G. RÖMER and E. RENNER 1974, P. G. GROENEN et al. 1976, M. GODEFROOT et al. 1981). Some of the equipment shown in figure 27 is taken from the review by J. M. H. BEMELMANS (1981). In a systematic comparative test of different isolation procedures, the distillation/extraction method was found to be most suitable for the isolation of the unchanged aroma spectrum from grapefruit juice (A. J. NUNEZ 1984).

The removal of the extractant without loss of aroma components is sometimes difficult. There are fewer problems during the early stages of distillation. Later on, long distillation times or partial losses of the most volatile components must be considered. Table 17 demonstrates this problem. Depending on the boiling point of the solvent the recovery rates change, and are influenced particularly by the final volume of the concentrated extract. One of the consequences is a changed aroma spectrum, because the losses are higher for low-boiling aroma components.

According to figure 26, which reviews the usage of various isolation and concentration techniques, distillation and extraction represent by far the highest proportion of all procedures summarized. Since the first review by C. WEURMAN (1969) the situation has not changed very much, except that adsorption techniques have overtaken chemical binding as an isolation method. Chemical binding is very appropriate and more commonly used today for thin layer chromatography (see **2.2.3.**). Classes of odorous components are often separated in this way from foods or food extracts.

Sorption methods which occurred with relatively low frequency in WEURMAN's 1969 review have become of increasing interest in recent years. Besides charcoal (J. H. DHONT and C. WEURMAN 1960, R. G. CLARK and D. A. CRONIN 1975, H. SUGISAWA et al. 1984),

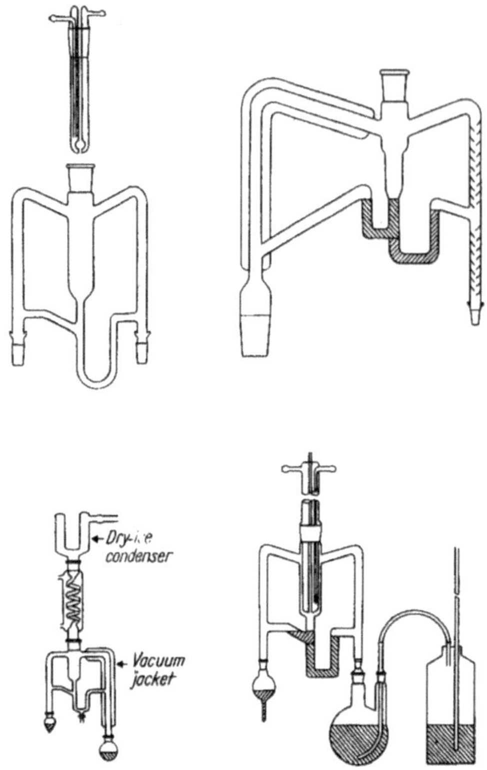

Fig. 27

Various Likens-Nickerson devices for simultaneous distillation/extraction of volatiles
(J. M. H. Bemelmans in H. Maarse and R. Belz, Isolation, separation and identification of volatile compounds in aroma research, Akademie-Verlag Berlin 1981, p. 22)

top left: Original apparatus as designed by Likens and Nickerson (26)
top right: Modification after Groenen et al. (40), suitable for extraction with 'heavy' solvents; solvent arm with Vigreux identations
bottom left: Modification after Maarse and Kepner (112) with vacuum jacket to minimize premature condensation and dry-ice condensor to reduce volatilization losses
bottom right: Modification after Römer and Renner (39); steam is generated in a separate vessel.

polymer adsorbents of synthetic character have been used with success. Among the frequently used polymers are p-2.6-diphenyl-phenylene oxide (Tenax GC) and gas chromatographic phases of the 'Chromosorb' and 'Porapak' type (S. Adam 1984).
In 1969 R. E. Kepner et al. published a review of the few articles in the field of freeze concentration. Using the equipment described by J. Shapiro (1961, 1967) for nine tested aroma components, recovery rates of 92 to 98% could be achieved. Nevertheless, the usage of such techniques has not increased very much up to now. Also rarely applied are the so-called zone melting procedures, where part of a frozen solution is melted and transported through the rest of the frozen medium (M. T. Huckle 1966, G. J. Arkenbout 1972).
At present it is safe to say that there seems to be no concentration procedure which avoids all losses in aroma. Thus in most cases we should take into consideration that the chromatogram registered after the separation step may be somewhat changed by the isolation and concentration technique (see for example N. Paillard et al. 1970).

2.2.3. Thin layer chromatography

In contrast to most of the other chromatographic techniques, thin layer chromatography is a very simple, fast and inexpensive method which works with easy-to-handle, uncomplicated equipment. E. Stahl (1956) first introduced tools for improving thin layer chromatographic techniques, and subsequently there were many inventions which promoted this type of analytical research in the aroma field (see the review of J. H. Dhont 1981). Examples of progress using special techniques and procedures are

— the separation of chemical classes of volatile aroma components by the formation of non-volatile, insoluble derivatives which can be easily pre-separated from the whole flavour complex (D. P. Schwartz and O. W. Parks 1963, G. Urbach 1969),
— the direct identification of aroma components by means of corrected R_F-values (D. S. Galanos and V. M. Kapoulas 1964, J. Casparic 1968, J. H. Dhont et al. 1970, 1972, J. H. Dhont 1981),
— the use of gas chromatographic separation techniques as an identification tool (H. W. H. Schmidt 1967, J. H. Dhont 1981),

Table 17
Influence of end volume and distillation time on the recovery rate of different aroma components (W. RÖDEL et al. 1971)

Aroma compound	B.P	Recovery rate with end volumina and distillation times of			
	(°C)	0.02 ml/2 h	0.4 ml/2 h	1 ml/2 h	1 ml/3.3 h
Butanol-1	75	7	9	36	75
Propanol-1	97	8	21	56	100
Pentanone-2	102	7	25	63	99
Ethyl-propionate	99	12	30	65	99
Butanol-2	108	14	32	78	
Hexanal	128	24	45	—	96
Isoamyl acetate	142	25	51	75	100
Heptanone-2	151	32	53	74	100

— the use of thin layer separation techniques of high efficacy such as two-dimensional separation, impregnation of stationary phases and reversed-phase chromatography or drum chromatography (J. H. DHONT 1981, D. L. SAUNDERS and R. L. SNYDER 1970) or
— the adsorption of selected volatiles on thin layer carriers (J. H. DHONT 1970).

Direct combination of gas chromatographic and thin layer chromatographic techniques were considered quite early (J. JANAK 1963, R. KAISER 1964, M. WINTER et al. 1967, J. H. DHONT 1981). By a special technique, the volatiles leaving the gas chromatographic column are directly adsorbed on the surface of thin layer plates. In this case the distance between the end of the column and the thin layer should not exceed 0.5 mm. During the gas chromatographic separation the plate is moved synchronously. M. WINTER et al. (1967) registered by means of this technique that 60 single components were hidden under 10 gas chromatographic peaks.

Figure 58 dealing with the separation of 2.4-dinitrophenylhydrazones on p. 86 demonstrates an example of the use of thin layer chromatography in bread aroma research.

2.2.4. Gas chromatography

The decisive breakthrough in aroma analysis was closely related to the introduction of gas chromatography. Compared to other chromatographic techniques, the advantages of this type of chromatography are obvious. They can be summarized as follows:

— a high separation effect,
— more variation of the separation conditions,
— the very low detection limit, which today in most cases is as low as our human sense organs,
— a relatively low time consumption,
— a good reproducibility of the analysis,
— a high degree of significance.

At present, gas chromatographic separation leads the field in aroma analysis and the other chromatographic methods have the character of supporting tools only. Within less than two decades, gas chromatographic equipment has been developed enormously (P. SCHREIER 1984). This trend has been encouraged by special techniques such as separation under programmed temperature regimes, the use of capillary high-resolution columns (W. JENNINGS and G. TAKEOKA 1984, T. SHIBAMOTO 1984), and the development of new detector systems (S. VAN STRATEN 1981). Moreover, the direct coupling of gas chromatography with identification methods such as mass spectrometry or other spectrometric methods has encouraged the rapid development and widespread introduction of gas chromatography (G. SCHOMBURG et al. 1984). Finally, the application of computers for detailed mathematical evaluation has also contributed to the growing importance of gas chromato-

graphy. The direct coupling of a gas chromatograph with a mass spectrometer and a computer can provide not only efficient identification of components but also quantification and data storing. It should be mentioned here, however, that parallel with the rapid development of chromatographic and spectrometric equipment there has been an exponential upward trend of the price of the apparatus.
Figure 28 tries to summarize some of the identification techniques used in aroma analysis. The isolation and concentration step is followed by gas chromatographic separation. In order to elucidate the complexity of the system, identification aids can help in some cases. Such tools are specific gas chromatographic detectors (detection of sulphur or nitrogen compounds), the use of reaction gas chromatography or the 'sniffing technique' of sensory evaluation of the effluents leaving the gas chromatographic column (see figures 33 and 34). For direct identification the combination with mass spectrometry is the method most widely used at present.
Gas chromatograms of naturally occurring food aromas generally deliver a very complex picture. A simplified view is possible in only two cases; first, if a direct injection is carried out, and second, in the case of the head space analysis discussed later on. But normally we can say that the more highly developed the resolution technique, the more complex the gas chromatograms appear. Figures 29 to 31 demonstrate this by using some selected examples.
Figure 29 shows a chromatogram of a brandy which was directly injected into the gas chromatograph without the isolation and concentration step. These time-consuming steps, which are normally involved in gas chromatographic analysis, can be avoided only when the concentration of the main volatile constituents is high. Modern methods in the field of direct injection techniques use micropacked capillary columns with thick films (G. REGLERO et al. 1986).
Isolation by extractive methods followed by a concentration step previous to gas chromatographic separation indicates a much higher solution of such complicated mixtures of volatiles. By this technique H. SUOMALAINEN and L. NYKÄNEN (1970) identified with packed columns 37 aroma components in whiskies. As shown in figure 30, the resolution is high enough to answer the questions which are of practical interest: differences in the composition of the three main whisky types.
The resolution, however, can be increased much more by the use of capillary columns. Figure 31 demonstrates a significantly higher number of aroma components for a rum extract which does not depend on the product

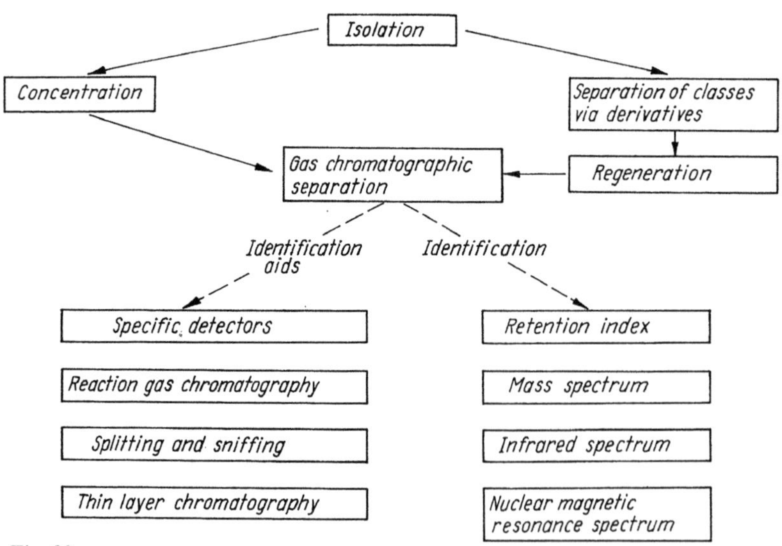

Fig. 28
Steps and tools of gas chromatoagraphic identification

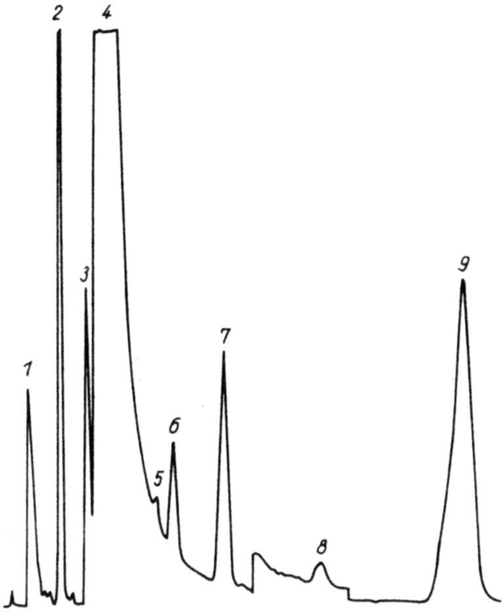

Fig. 29
Gas chromatogram of the dominating aroma components in rum using direct injection technique and separation on packed columns (M. ACKERMANN 1986)
1 = Ethanal
3 = Ethyl acetate
4 = Ethanol
5 = Ethyl butyrate
6 = 2-Methylpropanol
7 = 2-Methylbutanol + 3-Methylbutanol

but on the resolution technique. As early as 1966, H. MAARSE and M. C. TEN NOEVER DE BRAUW identified 47 out of a total of 101 separated substances. By means of modern capillary column techniques, much more compounds can sometimes be separated within one gas chromatographic step.

The special problems of high resolution techniques, however, should not be overlooked. Thus the separation effects published in the literature are often not easily reproducible (K. GROB 1974). Coupled gas chromatography and mass spectrometry are a very modern tool for the analyst. However, optimal results depend not only on a clear separation of the substance under test but also on the application of modern mathematical evaluation or computer techniques (M. NOVOTNY et al. 1974).

The comparison of **retention data** which are specific for every substance is a long-used effective method for evaluating the identity of the separated gas chromatographic peaks. The speed with which volatile components pass through the chromatographic column depends on the material, length and diameter of the column, as well as on the type and size of the stationary phase. In addition, many physical factors influence the passage, for example the temperature regime during the separation or the gas stream with its transport effect. However, under standard conditions homologous series of special classes of aroma components show a typical regular behaviour with respect to retention time. Testing this property on columns of different polarity may be useful as a tool for identification (see for example R. V. GOLOVNJA and V. URALETS 1968, 1971). From knowledge of the behaviour of the whole series of the class of compounds concerned, it becomes possible to calculate the identity of the component. Alkane homologues are normally used for this purpose. Figure 32 demonstrates as an example the identification of sulphur components within the aroma complex of meat products after separation by mercuric salts and regeneration. So-called KOVATS indices (E. KOVATS 1958, A. WEHRLI and E. KOVATS 1959) are calculated and used. At the top of the figure the chromatogram of the hydrocarbon series is reproduced, after separation under the same gas chromatographic conditions.

One of the main problems of the identification via KOVATS indices is the high number of compounds present in most gas chromatograms. By a pre-separation of single classes of aroma compounds the complicated chromatogram can be somewhat simplified (see figure 32). A further advantage of such

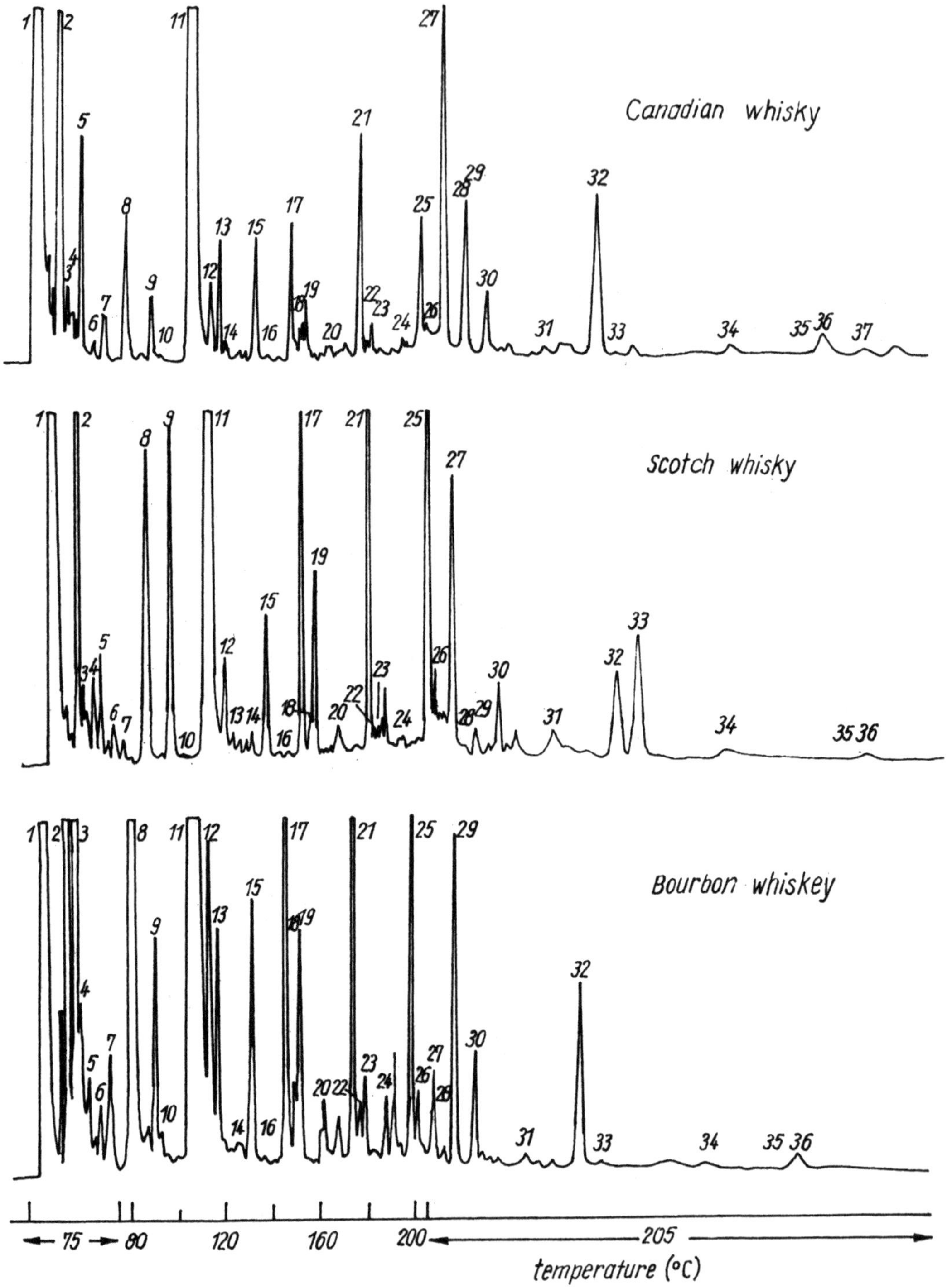

Fig. 30
Gas chromatogram of whisky extracts from different sources separated on packed columns
(H. SUOMALAINEN and L. NYKÄNEN 1970)

derivatization procedures is the precipitation or chemical binding of the frequently instable and volatile aroma components into stable derivatives which in most cases are solid and non-volatile. Thus they can often be separated by simple chemical operations. In this way it is possible, for example, to transform

- acids, amines and other bases into their stable and water-soluble salts,
- carbonyl compounds into non-volatile, intensily coloured 2.4-dinitrophenylhydrazones or
- sulphur compounds into non-volatile and stable mercuric precipitates.

R. V. Golovnja (1982), R. V. Golovnja and V. Uralets (1968, 1971) and V. Garbusov et al. (1976) used such derivatization procedures for the identification of many aroma compounds in meat, fish, milk products and bread.

Another important tool for identification of aroma components after gas chromatographic separation in the **sniffing technique** (splitting technique) already mentioned. In this method the gas stream leaving the chromatographic column is split into two parts at a known ratio. One part of the effluent loaded with the separated volatiles is led to the detector in the normal way, whereas another part leaves the apparatus through an open hole and can be tested by smelling.

A special 'sniffing' technique is used. Because of the low volume flow and mass transfer, humidified air is sometimes added to the gas stream, especially in the case of capillary columns. The method provides additional information about the sensory quality and odour intensity of each peak registered simultaneously.

The dependence of the sensory effect on the concentration of aroma compounds sometimes causes additional problems. Figure 33 demonstrates a typical chromatogram taken from the work of E. von Sydow's group on bilberries (1975), including some mass spectrometric identification data. Of the numerous publications in this field, that of T. E. Acree and J. Barnard (1984) dealing with apple aroma and methyl jasmonate isomers will be mentioned as an example.

Another identification aid is the use of specific group reagents brought into contact with the aroma complex, mostly used before the separation step. As an example, the so-called **reaction gas chromatography** uses a reaction column containing a group reagent. Such a column can be installed on the top of the column, or between column and detector. In this way special peaks are eliminated from the system by derivatization. By means of such a 'subtractive technique' functional groups of the peak substance can be recognized. This can facilitate the identification in many cases. Reactions like catalytic hydrogenation, esterification, ester hydrolysis, neutralization, ether formation or ozonolysis displace peaks

◀ 1 = solvent
2 = ethyl acetate
3 = ethanol
4 = ethyl propionate
5 = unknown
6 = isobutyl acetate
7 = ethyl butyrate
8 = 2-methylpropanol
9 = isoamyl acetate
10 = butanol
11 = isoamylalcohol
12 = ethyl caproate
13 = isoamyl isobutyrate
14 = isoamyl isovalerianate
15 = hexanol + ethyl lactate
16 = isoamyl valerianate
17 = ethyl caprylate
18 = isoamyl caproate
19 = furfural
20 = octanol + ethyl nonanoate
21 = ethyl caprate
22 = isoamyl caprylate
23 = ethyl-9-decenoate
24 = decanol + ethyl undecanoate
25 = β-phenylethyl acetate + ethyl laurate
26 = isoamyl caprate
27 = β-phenylethanol
28 = dodecanol
29 = β-methyloctalactone
30 = ethyl myristate
31 = tetradecanol
32 = ethyl palmitate
33 = ethyl palmitoleate
34 = hexadecanol
35 = ethyl stearate
36 = ethyl oleate
37 = ethyl linoleate

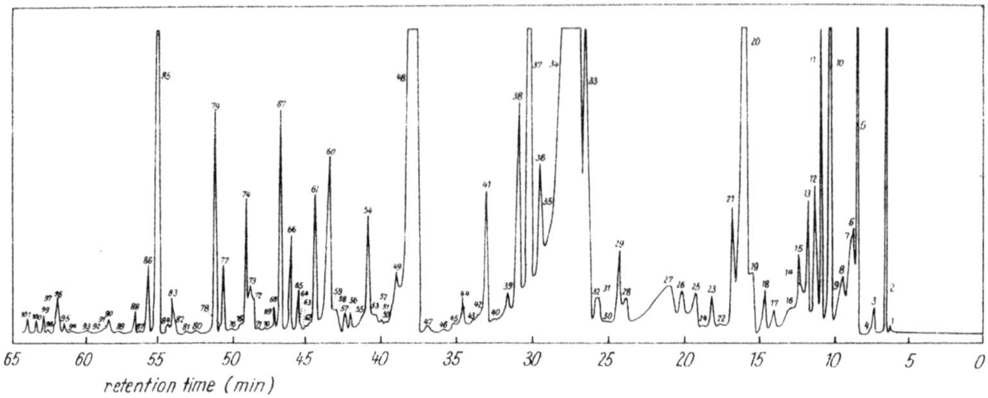

Fig. 31
Gas chromatogram of rum extract separated on a capillary column
(H. MAARSE and M. C. TEN NOEVER DE BRAUW (1975)

1, 2 = solvents
3 = ethyl formiate
4 = 2-methylpropanal
5 = ethyl acetate
6 = ethanol
7 = diethoxymethane
8 = 2-methylbutanol
10 = 1,1-diethoxyethane
11 = ethyl propionate
12 = propyl acetate + propanol
13 = butanol-2
15 = ethyl isobutyrate
17 = 1,1-diethoxypropane
19 = isobutanol
20 = ethyl butanoate
21 = 1,1-diethoxybutane
23 = 1-ethoxy-1-(2-methylpropoxy)-ethane
25 = ethyl 2-methylbutanoate
26 = ethyl 3-methylbutanoate
27 = butanol
28 = 1-ethoxy-1-butocyethane
29 = 3-methylbutyl acetate
31 = propyl butanoate
32 = 2-methyl-furanidon-3
33 = ethyl pentanoate
34 = 3-methylbutanol
35 = 1,1-diethoxy-2-methylbutane
36 = 1-ethoxy-1-(2-methylbutoxy)-ethane
37 = 1-ethoxy-1-(3-methylbutoxy)-ethane
48 = ethyl hexanoate + furfural
49 = 1-propoxy-1-(3-methylbutoxy)-ethane
54 = 1-ethoxy-1-(3-methylbutoxy)-butane
57 = 1-butoxy-1-pentoxyethane
60 = 2-acetylfurane
66 = propyl hexanoate
67 = ethyl heptanoate + benzaldehyde
68 = 1,1-butoxy-pentoxyethane
72 = 1,1-ethoxy-pentoxybutane
73 = 5-methylfurfural
74 = 1-ethoxy-1-pentoxypentane
77 = 1-(2-methylbutoxy)-1-(3-methylbutoxy)-ethane
79 = 1,1-di-(3-methylbutoxy)-ethane
83 = 2-methylpropyl hexanoate
85 = ethyl octanoate
90 = 2-methylbutyl hexanoate
95 = ethyl benzoate
96 = diethyl succinate
101 = ethyl nonanoate

from the chromatogram, thus proving that they belong to particular chemical classes.

The main problem of reaction gas chromatography, as with other derivatization procedures, is an incomplete and/or nonspecific reaction. Too short, or too long reaction times and a limited specifity of the chemicals used with respect to the traces of aroma compounds, may play a role. Thus a clear classification is frequently not easy to attain.

At present, however, the identification tool most often used is the **combination** of gas chromatography **with spectrometric techniques**. Mass spectrometry is the most favoured method for direct coupling with the gas chromatograph. Modern equipment is designed to have suitable sensitivity for both techniques, which facilitates coupling. Compared with mass spectrometry the other spectrometric methods require more material for a clear signal than the very sensitive gas chromatographic detectors. Enrichment of separated gas chromatographic peaks can be realized by so-called 'preparative gas chromatography'. Preparative gas chromatographs are equipped with a dosage system. According

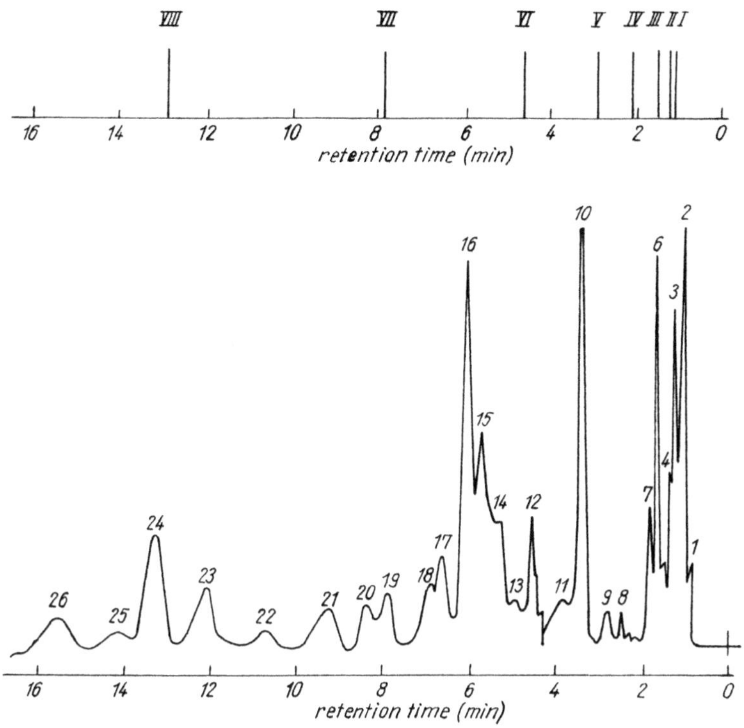

Fig. 32
Gas chromatogram of sulphur compounds isolated from natural boiled beef via mercuric salt adducts
(V. GARBUSOV et al. 1976)
I—VIII = retention time of n-alkane homologues with 5 to 12 C-atoms
3 = 1,2-propylene sulphide
4 = thiophene
7 = 2-methylthiophene + dimethyl sulphide
9 = furfuryl thiol
12 = 3-tert. butylthiophene + thiophenol + 1,3-dithiolane
13 = diisopropyl sulphide
14 = heptyl thiol
16 = 1,4-butanedithiol
18 = sec. octyl thiol + dibutyl sulphide
20 = di-tert. butyl sulphide

to the required procedure, the separation process is repeated several times. After the peak in question has passed the column and is isolated in a special device, separation stops. Then the apparatus is cooled back to the starting temperature, and a new injection follows. The small amounts of aroma compounds separated in one gas chromatographic separation, however, often require many repeated injections: time therefore becomes the limiting factor.

Special detectors are another aid for identification. In this way the number of substances indicated is reduced to those which are sensitive to the detector signal. Figure 34 shows an example of the application of a sulphur-specific detector in the identification of aroma components in boiled chicken. The gas stream leaving the gas chromatographic column after the separation process is split into two parts, one of which is detected by the normally used flame ionization detector (FID) and the other by a SO_2-sensitive detector (FPD). The latter indicates all substances which contain sulphur atoms. The parallel running of both chromatograms makes it possible to

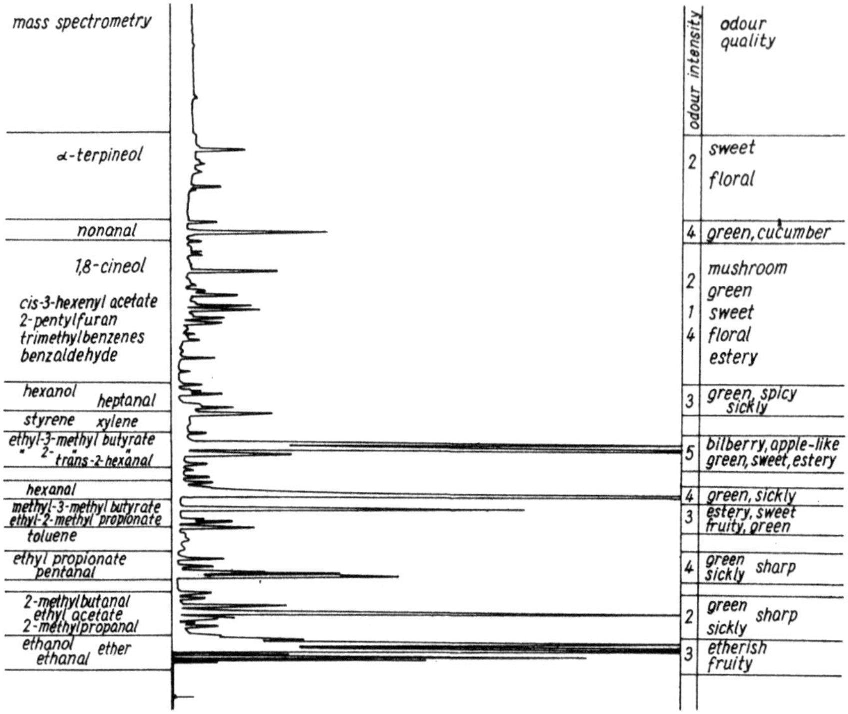

Fig. 33
Gas chromatogram of concentrated head space volatiles in bilberry juice including mass spectrometric and sensory results
(E. VON SYDOW 1975)

recognize which of the components are sulphur-containing. Additionally, the chromatogram shows that the two detectors used differ in sensitivity: the sulphur components are indicated in a higher amount as in the case of the FID.

One of the most important variants of a gas chromatographic examination is the direct analysis of the so-called 'head space' over the aromatic food medium. In the simplest form **head space analysis** is performed by enclosing the sample in a gas-tight vessel, waiting until equilibration is reached and finally taking a measured part of the gas in the 'head space' for analysis, which is then injected into the chromatographic column. A modification of this kind of analysis is the trapping of volatiles with a gas stream and their condensation or sorption in special traps ('cumulative collection', see J. SCHAEFER 1981). Here the total content of aroma components is not included into the analysis: only the volatilized part present in the gas phase over the medium. We can differentiate between four groups of application:

- the direct use of liquid and frequently aqueous samples, such as beverages, essential oils or spirits,
- the application on heterogenous liquid and solid samples, such as dairy or fruit products,
- the use of solids which can be converted into an approximately homogenous liquid, such as instant coffee, instant fruit juices or instant cocoa powders,
- solid foodstuffs which emanate volatiles after crushing into a mixture of solids and an aqueous phase.

One of the main advantages of head space analysis is the reduction of the analysis on the volatilized part: that is that part which can be directly smelled. Another advantage is the

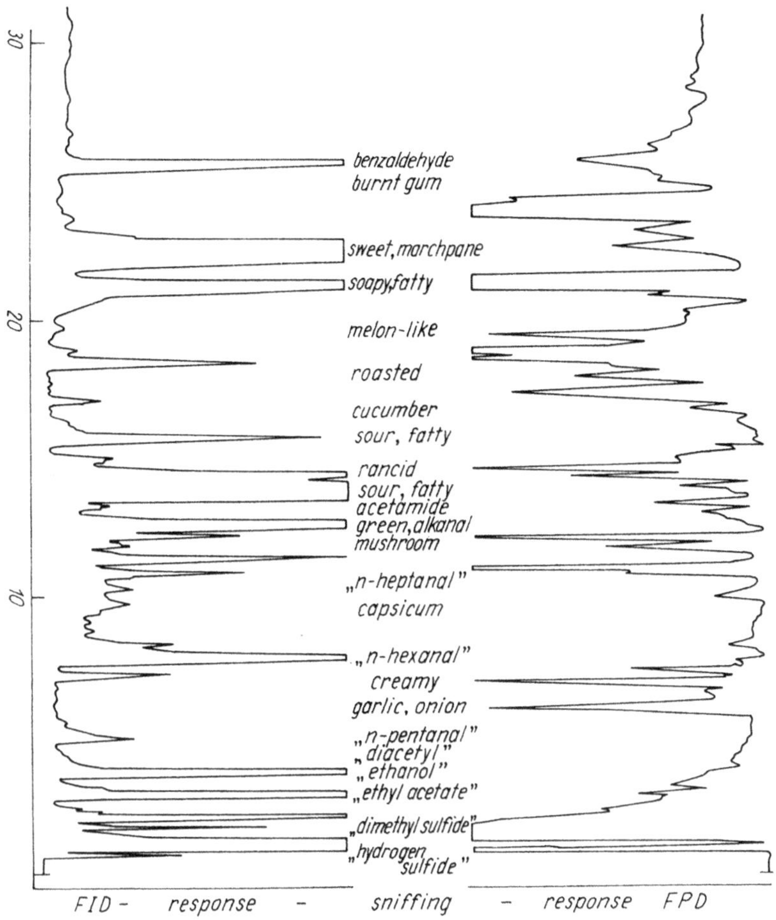

Fig. 34
Gas chromatographic profiles of boiled chicken aroma volatiles detected by means of parallel use of a flame ionization detector (FID) and a flame photometric detector (FPD) and peak sniffing
(R. SCHRÖDTER 1985)

lack of substance losses and/or artefact formation in the course of isolation and concentration procedures. But the most important advantage of this method is the simplicity of sampling, which saves a great deal of time. A possible problem is relatively poor reproducibility, but this depends, however, on the be tested. A strict control of the temperature regime is necessary. The very small amount of volatiles present is another problem, which may be overcome by modern methods and equipment (G. OGLE 1983, W. BOLAND et al. 1984).

For more than two decades, gas chromatographic separation and determination have been central to the aroma research literature. Modern equipment and techniques, and many experimental variants, have made accessible a wide field of special methods (S. VAN STRATEN 1981). Here we have been able to demonstrate only a very few examples.

Recently, a new trend has been observed in gas chromatographic aroma research. Formerly, as a consequence of losses during the isolation and concentration procedures, qualitative analysis and identification received the greatest attention. Practical problems, however, often include quantitative aspects: for example flavour quality control and problems of aroma formation or stabilization, as well as flavouring production by use of fermentative or thermic processes, need quantitative data. In the recently published literature there are examples of new approaches to quantified gas chromatography, for example, after direct injection of fruit essences by use of capillary columns (M. G. MOSHONAS and P. E. SHAW 1984). Recently, first summarized lists with quantitative data of food volatiles have been published (H. MAARSE and C. A. VISSCHER 1982—1986; L. NYKÄNEN and H. SUOMALAINEN 1983).

2.3. Spectrometric identification

Today, spectrometric methods provide the surest method of identification of aroma substances. Such methods generally use the excitement of the molecules by electromagnetic waves. From the absorption or emission stimulated under standardized conditions, conclusions with regard to the structure of the compound under test can be drawn.

Compared with the direct use of gas chromatographic retention data for identification purposes, mass or infrared spectrograms include far more data (R. BELZ 1981). Moreover, these methods are readily applicable for micro samples such as are available from the output of a gas chromatograph.

Even when the separated components leave the outlet of the gas chromatograph in submicrogram amounts, sensitivity and detection speed of the system show a similar efficacy to the record of the mass spectra. A direct coupling is therefore possible. Such combinations do not affect the separation effect of the gas chromatograph, and are now very popular in spite of the high price of the combined equipment (M. C. TEN NOEVER

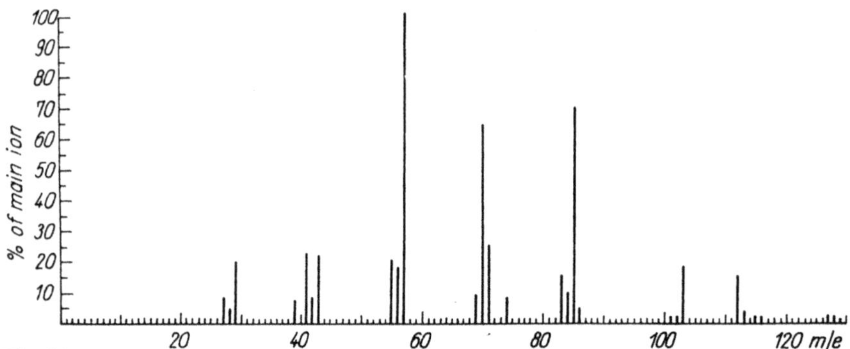

Fig. 35

Mass spectrum of hexyl 2-methylbutanoate
(R. SCHRÖDTER 1977)

Mass	% of B	Fragments	Mass	% of B	Fragments
27	9	C_2H_3	84	10	n. i.
28	5	CO, C_2H_4	85	69	C_4H_9CO, C_6H_{13}
29	20	C_2H_5	86	6	n. i.
39	8	C_3H_3	101	3	mol peak — 85
41	22	C_3H_5	102	2	mol peak — 84
42	77	C_3H_6	103	18	mol peak — 83
43	22	C_3H_7	112	16	mol peak — 74
55	20	C_4H_7, C_2H_3CO	113	3	n. i.
56	18	C_2H_4CO	115	2	mol peak — 71
57	100	C_4H_5, C_2H_5CO	116	2	mol peak — 70
69	9	C_3H_5CO	127	2	n. i.
70	64	C_2H_5CHCO, C_5H_{10}	128	2	n. i.
71	25	C_3H_7CO, C_5H_{11}	129	1	mol peak — 57
74	8	$CH_3CH{=}COHOH$	157	1	mol peak — 29
83	16	C_4H_7CO, C_6H_{11}	186	1	mol peak

n. i. = not identified

de Brauw and C. Brunner 1967, R. Teranishi et al. 1971, K. D. Schuy and D. H. Hunneman 1971, S. S. Chang 1973, M. C. Ten Noever de Brauw and C. van Ingen 1981).

In contrast to other spectrometric techniques, in which the substance under test remains unchanged, during mass spectrometry the molecules are split into fragments. The type and amount of the fragments characterize the original structure.

After volatilization of the test substance under high vacuum (10^{-6} Torr) the analysis starts with ionization, for example by heating a tungsten or rhenium wire (electron impact ionization) and acceleration of the emitted electrons to an anode. The bombardment of the molecules of the test sample by these electrons, with an energy of 20 to 100 eV, causes the splitting of some molecules into fragments or the ionization of the molecule itself. The pattern of the fragments is characteristic of the molecular structure.

The ions formed are positively charged, and accelerated by a negative tension. Finally they are separated by a magnetic or electric field according to their mass/energy ratio. A line spectrum is obtained (see figures 35 and 36). Of special importance is not only the fragmentation pattern but also the so-called *mol peak*, which is identical with the highest m/e ratio. It is caused by the ion with an unchanged molecular mass, and thus gives an important reference for the identification of the substance.

The appearance of fragments follows special rules, the knowledge of which is used in determining the probable structure. Test spectra for comparison are useful here. For inter-

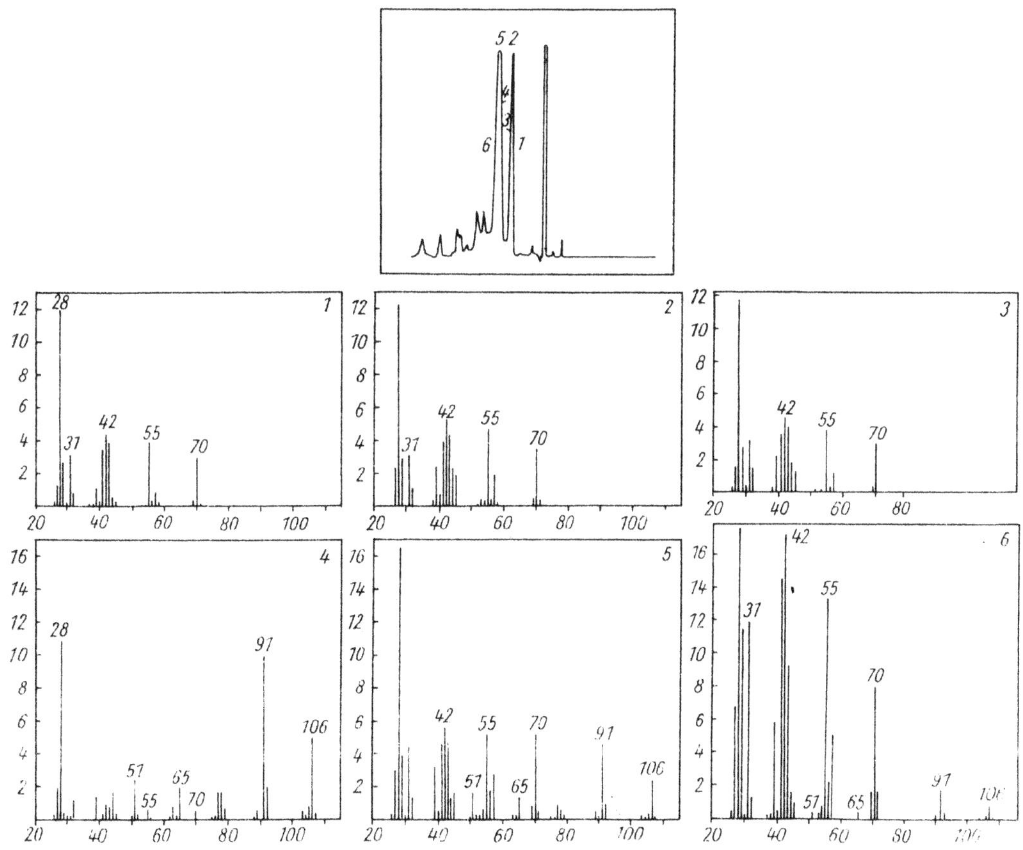

Fig. 36
Capillary gas chromatogram and mass spectra of a mixture of sec.-butanol, isoamyl alcohol, ethyl benzene and amyl alcohol
(M. C. ten Noever de Brauw and C. van Ingen 1981)

pretation of mass spectra, however, experience in this field as well as chemical knowledge is essential.

We can summarize the main advantages of the mass spectral identification of aroma components as follows:

- very high sensitivity
- possibility of direct coupling with gas chromatographic equipment
- a relativ simple relation between the structure of the substance under test and its mass spectrum
- possibility of recognizing the molecular mass directly
- coupling with a computer for sampling and storing the large amount of data produced.

The on-line coupling of gas chromatograph and mass spectrometer, as well as the additional combination with a computer, result in a very high efficacy. All the data from a complicated gas chromatogram, such as is typical for most aroma complexes, can be registered, stored, calculated and identified. So-called *interfaces* have to be used, because of the pressure differences between the gas chromatograph and the mass spectrometer which work under vacuum conditions, and for removing most of the carrier gas from the traces of substances under test.

Figure 37 shows the main outlines of a combined gas chromatographic/mass spectrometric examination taken from earlier work of E. J. MULDERS and C. WEURMAN (1970). In the head space over wheaten bread, the aroma components present were enriched stepwise using deep-cooled pre-columns. After three runs the gas chromatogram is clear enough and suitable for mass spectrometric identification. This reminds us of one of the main requirements of mass spectrometric analysis, that means absolute purity of the substances tested. Nowadays, however, modern gas chromatographic separation techniques have become so diversified that practically any organic compound can be isolated

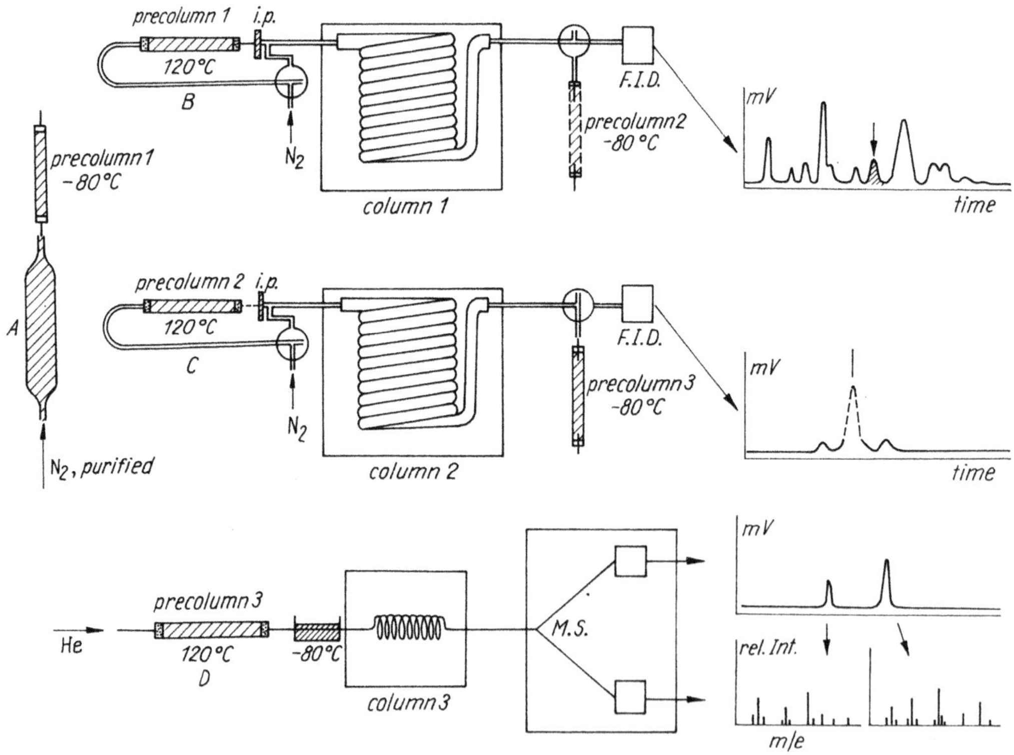

Fig. 37
Mass spectrometric identification of gas chromatographically separated, pre-enriched head space samples (E. J. MULDERS and C. WEURMAN 1970)

and identified by its mass spectrum (M. C. ten Noever de Brauw and C. van Ingen 1981).
If we know what type of test substance we are dealing with, the mass spectrometer can also be used as a detector. By means of gas chromatographic retention data and mass spectrometric values, formation mechanisms of aroma compounds from percursors have been evaluated in this way (H. Boelens et al. 1974). T. Hirvi (1983) applied a mass fragmentographic technique for the quantitative analysis of 17 main components in strawberries. The concentrations present in different varieties were quantified by determining two monitored ions of each compound separated by gas chromatographic analysis.
Among the other spectrometric identification techniques currently used infrared (IR) spectroscopy and nuclear magnetic resonance spectroscopy (NMR) are of some importance in aroma research. Compared with mass spectrometry, both principles require higher amounts of the test substance (NMR spectroscopy: 10^{-3}—10^{-2} g; IR spectroscopy: 10^{-6}—10^{-4} g; mass spectrometry: 10^{-9}—10^{-7} g). This causes additional problems of pre-separation, concentration and purification or, for example, enrichment by preparative gas chromatography. The direct coupling of these spectrometers with the gas chromatograph is possible, but depends on several special pieces of apparatus for increasing sensitivity and reducing recording time of the complete spectrum without band distortion (see R. Belz 1981).
Figure 38 shows the IR spectrum of an aroma component with typical bands. The interpretation of such spectra roughly follows two steps: the first is the determination of functional groups within the molecule. The bands

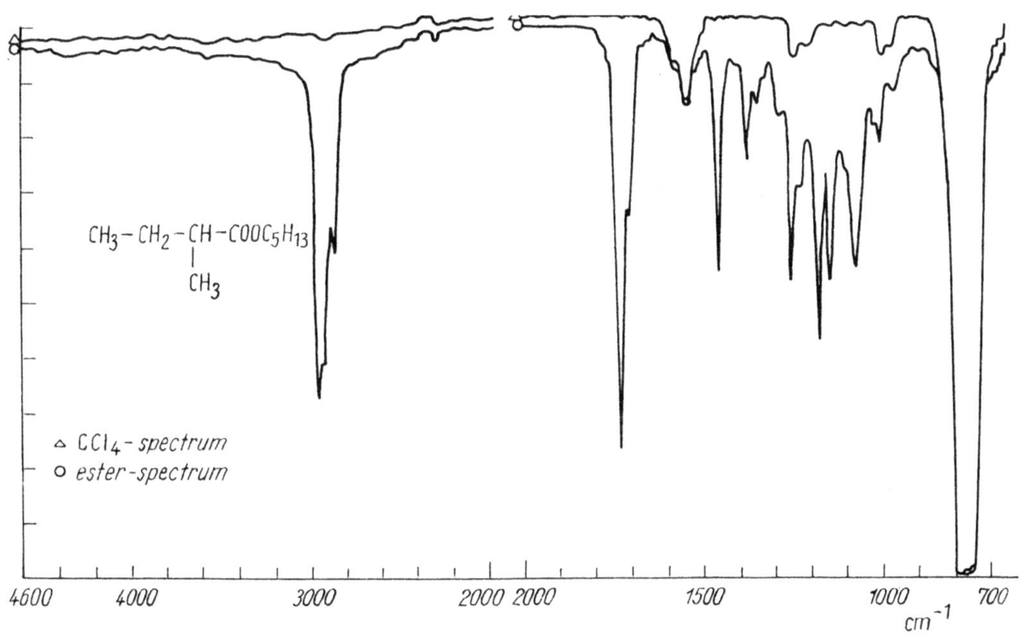

Fig. 38
Infrared spectrum of hexyl 2-methylbutanoate
(R. Schrödter 1977; 10% ester in CCL_4; NaCl cuvette 0.03 mm)
Characteristic bands (cm^{-1}):

CH, CH_2, CH_3/CH valence oscillation	2850
	2910
	2950
C — CH_3	1380
	1460
C = 0	1730
C — 0 — C asymmetric valence oscillation	1180
C — 0 — C symmetric valence oscillation	1080

of an IR spectrum provide some information about the type and position of functional groups. The second step is normally the comparison of the 'fingerprint picture' of the test substance with reference spectra (R. BELZ 1981). Such spectra have been published for many aroma components (W. W. MORRIS 1973, M. HORAK et al. 1960/63, J.-A. WENNIGER et al. 1970).

2.4. Food aroma components

2.4.1. Occurrence in food and beverages

Of the three million organic substances currently known, most are odour active: especially those with components of low to medium molecular mass. If the molecular size exceeds a certain limit, its volatility obviously becomes too low to transport to the odour receptor cells the number of molecules necessary to produce a signal.

Though nearly all volatile organic substances produce odour effects (H. G. MAIER 1970), only a small proportion of them cause pleasant sensations and contribute to food aromas. There are many other compounds which we associate with other domains of life. Examples are the smell of flowers, of wooden material, of hay and soil or (within the technical area) the odours of plastics, of leather, of mineral oils, petrol products or burnt gum. Many volatile organic compounds produce odours which must be considered as 'off-flavours' in terms of food aroma.

In foods, beverages and stimulants like coffee, tea or tobacco, nearly 4500 aroma components have so far been identified (H. MAARSE 1981, 1985, S. VAN STRATEN and F. J. G. DE VRIJER 1977 and supplements). Our knowledge in this field is increasing rapidly, in parallel with the development of analytical techniques for volatile trace compounds. The total number of aroma components is expected to be between 5.000 and 10.000 (F[1]) RIJKENS and H. BOELENS 1975). Among the compounds identified esters predominate with approximately 450, followed by carbonyl compounds and terpenes each with 300 known representatives.

The qualitative analysis of single food aromas was the first target when gas chromatography was introduced into aroma research nearly three decades ago. The number of identified compounds rapidly increased. But with increasing knowledge it became more and more obvious that the original concept of a limited number of aroma compounds within one food aroma is wrong. Simple aroma spectra are rather rare, and are to be found mainly in some spices like vanilla, bitter almond, cinnamon or peppermint. Here special 'key substances' dominate within the aroma complex. This is the exception, however, and not the rule. Nevertheless it was these properties of certain herbs, spices and fruits which caused their early use for seasoning purposes. In this case one aroma component is present in such a high concentration that the addition of very small quantities in cooking transmits the typical aroma to the meal.

However, in most cases the aroma spectrum of foods is much more complicated. The lists of volatile compounds in food, first collected, summarized and published by C. WEURMAN (1963) and continued by S. VAN STRATEN and F. J. G. DE VRIJER (1977) and H. MAARSE (1985), are an excellent tool for all specialists dealing with aroma and flavouring problems because they summarize the nature-identical aroma components. The picture they provide is, however, confusing, and its complexity is continuously increasing. As shown in table 18, an ordering of these compounds by their chemical classes does little to improve matters. If the lists of aroma components mentioned are studied in detail, some additional information may be found. But a high number of esters identified in fruits, of sulphur compounds in meat products, or of pyrazines in cocoa or roasted coffee, for example, can only indicate the direction in which future research should be concentrated.

It is believed that generally aroma compounds do not contain halogen or phosphorus atoms[1]).

There are only a few exception to this rule, for example the identification of 2-acetyl-5-chloro-pyrrol in cocoa butter (C. T. HO et al. 1982) or halogenated alkyl and aryl components in the volatiles of cheese (J. P. DUMONT et al. 1974). Here, pollution effects should be considered.

Table 18
Number of aroma compounds identified in different foods and their distribution into chemical classes (S. VAN STRATEN and H. MAARSE 1983)

	Total[1]	Terpenes[2]	Hydro-carbons	Alcohols	Aldehydes	Ketones	Esters Lactones	Acetals	Furanes	Amines + other bases[3]	Pyra-zines[4]	Sulphur compounds
Apple	329	11	34	42	25	14	129	19	3	4	—	3
Strawberry	351	13	29	56	18	20	139	20	8	1	—	7
Banana	228	1	—	48	10	14	100	—	—	8	—	—
Orange	245	77	51	53	35	14	53	2	1	—	—	5
Asparagus	137	—	—	25	15	16	5	—	7	5	17	32
Cabbage	96	—	—	9	24	7		—	1	14	—	37
Celery	142	59	25	23	18	11	18	—	1	—	1	2
Onion	85	1	1	5	8	3	—	—	2	1	1	61
Potato	126	9	24	22	22	11	1	1	6	5	4	23
Potato, baked	252	7	52	22	35	18	17	3	9	6	46	8
Tomato	324	20	31	46	54	34	43	3	21	10	5	20
Peanut, roasted	365	7	51	27	48	27	19	1	18	17	54	38
White bread	280	1	11	18	40	33	25	4	20	37	47	18
Cheddar cheese	161	4	33	12	12	14	29	—	—	14	—	6
Swiss cheese	120	2	16	8	10	12	20	—	—	8	6	2
Beef, heated	611	2	85	46	54	54	45	2	36	31	67	139
Cocoa	387	21	44	27	24	29	62	6	16	19	84	18
Coffee	628	4	66	18	25	62	38	1	83	81	96	91
Beer	439	29	14	56	39	36	96	1	12	37	30	32
Wine	493	12	24	57	28	19	189	8	12	23	1	14
Cinnamon	89	35	24	10	9	4	9	—	—	—	—	—
Pepper	110	74	38	27	3	11	10	—	—	8	—	—

1) including additional groups of volatiles
2) terpene hydrocarbons, terpene alcohols, terpene aldehydes and terpene ketones; this group is registered additionally under the four chemical classes
3) including pyrroles
4) including pyridines and piperidines

This may be explained by the fact that halogenic substances produce unpalatable odour notes. Phosphorus is normally present in nature in the form of phosphates: such substances have very low volatility, which restricts odour perception.
The number of aroma compounds found in a food naturally indicates neither the aroma intensity nor the importance of the volatiles for flavour perception. Furthermore, the identification of an aroma component in a food product does not tell us whether this substance can actually contribute to aroma sensation in the concentration present. The fact that an identified component is known to have an odour in the pure state cannot be accepted as a proof of its participation in aroma or flavour of the food in question.
Consequently the search for 'key substances' within the aroma complex of foods was one of the earliest main aims of aroma research. W. G. JENNINGS and R. SEVENANTS (1964) were among the first who were able to prove the presence of key compounds in their aroma studies on Bartlett pears. They defined ethyl decadienoate, which they found to be of special importance for the aroma type of this fruit, as a 'character impact compound'. All other components identified in the pear, which obviously contribute much less to the aroma sensation, were characterized as 'contributory flavour compounds'.

Two decades after this definition, the number of known key substances within the aroma of foods has remained low, although there has

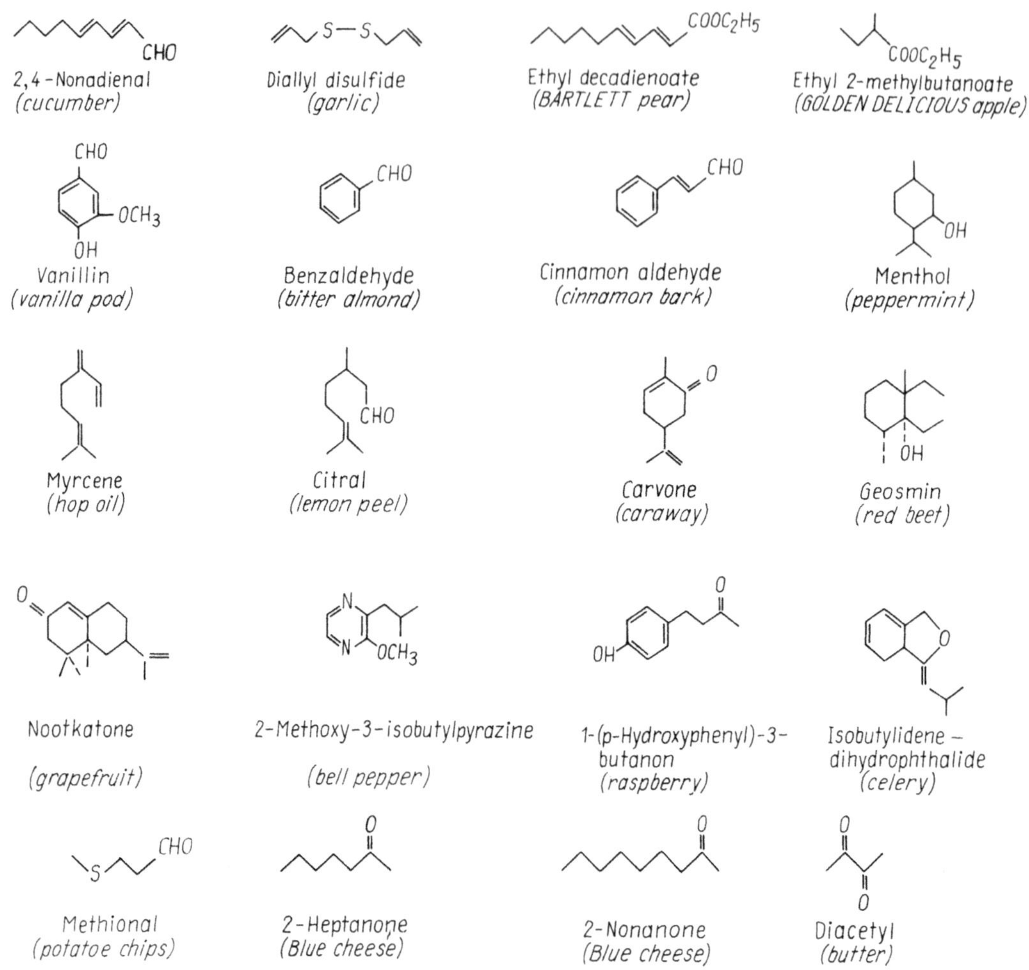

Fig. 39
Aroma components discussed as 'key components' within food flavours

been much progress in the aroma field. Some examples are summarized in figure 39. We can see that 'key components' belong to different chemical classes, and have generally been found in naturally occurring food flavours ('primary aroma components'). Spices, fruits and vegetables are the main sources. Methional in potatoe chips and 2-heptanone and 2-nonanone in Blue cheese are the exceptional examples of key components produced in the course of food processing ('secondary aroma components'). Considering the sensory properties of these key components, such substances can be defined as components which play a central role within a special aroma complex which is completed, however, by the presence of many other additional aroma compounds.

The gas chromatographic 'splitting technique' has frequently been used as a tool for looking for key substances within an aroma complex. The gas stream which leaves the gas chromatographic column is divided: one stream goes to the instrumental detector, while another one leaves the equipment and is evaluated by sniffing. In this way, the odour of each single peak registered by the detector can be checked.

In most cases, however, the characteristic top notes of an aroma cannot be found in a particular part of the gas chromatogram. This method has therefore given support to the idea

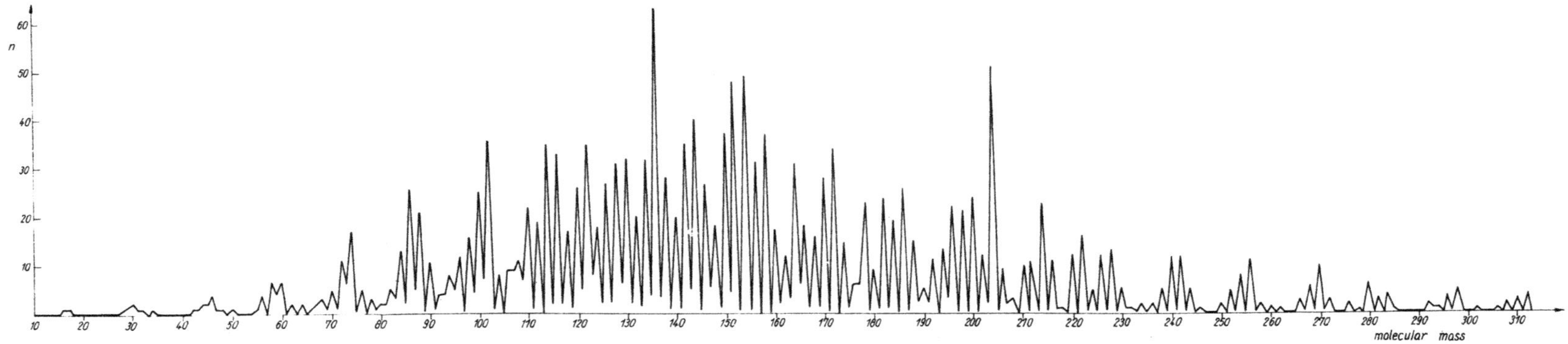

Fig. 40
Frequency of molecular masses of aroma compounds (2000 aroma compounds evaluated, round numbers)

that the aroma of most foods is caused not by single key components, but by the simultaneous reception of many odorous substances (O. G. VITZTHUM 1976). Beyond this conclusion is yet another hypothesis, F. V. KOSIKOWSKI and G. MOCQUOT, as early as 1958, assumed in their 'component balance theory' that for every aroma impression a special quantitative ratio of numerous compounds is necessary. The aroma of the single isolated substances may be totally different from the impression of the whole complex. Well-known examples are some sulphur compounds (simple thiols, furfuryl disulfide) present in coffee or meat, which produce unpleasant odour notes in the pure form but contribute to the positive aroma within the complex.

As long as the number of known aroma components increases by research on identification, the possibility cannot however, be excluded that 'key substances' will be found among those not yet identified.

Aroma compounds are characterized by two typical properties, i.e. high volatility and high aroma effectiveness. These attributes are related: only relatively small molecules are volatile enough for the necessary transport to the odour receptors.

Thus a relatively low molecular mass is one of the requirements for high aroma efficacy. J. EISENBRAND (1971) first noted that the molecular mass of aroma components does not exceed 300: he found the most usual value to be in the range of 100 to 150. From the summarized data of S. VAN STRATEN and F. J. G. DE VRIJER (1973) with 2.000 aroma components we find the distribution shown in figure 40. The values are rounded to whole numbers, each point represents the sum of two molecular masses.

From figure 40 it seems that molecular masses between 135 and 155 are most frequent. The upper limit is found to be near 300. Some of the high frequencies can be explained by the occurring isomers of terpenes in normal form (molecular mass = 135) and oxygenated form (molecular mass = 154), or the sesquiterpenes and diterpenes respectively. Aroma components of very high aroma efficacy have molecular masses below 200 (M. ROTHE and M. SPECHT 1976).

2.4.2. Aroma efficacy and threshold value

The problem of finding out the most important components within the very complicated aroma spectra of foods cannot be solved by quantitative analysis alone. The substances which predominate within the volatiles need not necessarily be identical with the most important ones. This fact is easily demonstrated by consideration of a simple example.

Let us compare the aqueous solutions of two well-known aroma components, vanillin and ethanol, each in a concentration of 0.1 % = 1.000 ppm. If we check these two solutions by sniffing or tasting in the mouth, we will register a very high difference in odour and flavour effect. While the vanillin solution produces a very intensive impression, the ethanol solution can only with difficulty be differentiated from pure water.

In contrast, however, a quantitative instrumental analysis would in both cases yield the same concentration value of 1.000 ppm. It is obvious from this example that we must avoid the erroneous conclusion that the same analytical value necessarily indicates the same aroma intensity.

Aroma components consequently differ with respect to their aroma efficacy. Practical conclusions can be draw from analytical data of aroma components only when this important property is also taken into consideration.

As aroma efficacy is a sensory property, a sensory method is required for its determination. But there are some physiological barriers which limit a direct evaluation: it is therefore rather difficult for untrained people to determine the intensity of different odour qualities. Human beings are equipped with a very good memory for special aroma nuances (as in qualitative sensory analysis), but can scarcely remember aroma intensities (as in quantitative analysis). Training, or reference substances are necessary (J. F. CAUL 1975).

The same is true for other sensory impressions. If we are well enough trained, we find it easy to define the quality of sounds. But this hearing, for example, within our musical memory is also limited, within our musical memory we find it much harder to quantify sounds.

The application of special differentiating test methods enables us to discriminate between a large series of various odour or taste intensities. K. HOPPE, using such tests for psychophysical studies, found 25 steps for sweetness differentiation (1983). Because of the difficulty of checking the aroma efficacy of aroma components directly, the indirect determination via threshold values was discussed early in the literature. In practice, a known concentration of the aroma compound to be tested is diluted with a neutral agent (water, air or some other medium) until the detection limit for human senses is reached. This so-called 'threshold value' indicates the minimum concentration which can be significantly differentiated from the diluting agent. By a systematic comparison of blank samples with solutions of different concentration, the perception limit can easily be determined. It is necessary, however, to work with an trained test panel, and to evaluate the results by statistical methods. Threshold value and aroma efficacy of aroma components are negatively correlated, so threshold data may be used as an index for this sensory property.

Two types of threshold value can be defined: the **detection threshold** ('stimulation threshold') characterizes the lowest concentration where a significant difference from the blank can be recognized. The term **recognition threshold** refers to the minimum concentration that permits recognition and definition of a changed flavour, compared with the dilutant. The determination of detection threshold values is easier and more practicable in most cases.

Threshold values of aroma components depend on the type of diluting agent used. On account of similarity to natural conditions, air or water are most frequently used. Threshold values in air can be obtained by an olfactometer, which allows not only an exact dosage but also the uniform distribution of the test sample in the gas phase and a reproducible supply to the human nose (see H. ZWAARDEMAKER 1921, M. STUIVER 1958, S. OUGH and H. STONE 1961, E. P. KÖSTER and O. WOUTERS 1970, S. VAN GEMERT 1981). Olfactometers have been most often used in the past for physiological tests, such as proof of changes in the odour sense organ or the odour centre of the brain; determination of fatigue effects after repeated irritation; or adaption to a special odour.

Gas chromatographic 'sniffing techniques' have also been used for threshold determination of volatiles in a gaseous atmosphere. Recently F. DRAWERT and N. CHRISTOPH (1984) described a standardized method for this type of determination in gas chromatographic effluents. The data were used for an interpretation of quantified head space analysis.

Threshold determination in a liquid, especially in an aqueous medium, seems to be easier. Again the threshold value depends on the solvent medium: in practice water and vegetable oil are frequently used, being naturally occurring media.

In the literature many procedures for threshold determination are recommended. The results, however, depend on the test method as well as on the test panel. Consequently, rather high differences can be registered by comparing the values available in the literature (see, for example, F. A. FAZZALARI 1978, N. BARYLKO-PIKIELNA et al. 1983). Table 19 reviews threshold data obtained by three different methods and groups:

- squeezing plastic bottles and smelling, (D. D. GUADAGNI et al. (1963)),
- smelling on open beakers, (E. J. MULDERS (1973)),
- tasting in the mouth (M. ROTHE et al. (1972)).

Threshold data are also influenced by the test temperature. The lowest concentration was found when threshold values were determined at 22 °C (R. FISCHER 1971).

According to table 19, large differences must be expected in aroma efficacy of different aroma compounds: they range over eight orders of magnitude.

As a general conclusion on the aroma efficacy of different substances, we are justified in saying that highly effective aroma components can be found among the sulphur and carbonyl

Table 19
Threshold válues of selected aroma components in dependance on the test method (detection threshold in ppm; water as diluting agent; 20 °C)

Test method	Squeezing[1] plastic bottles	Smelling[2] on beakers	Tasting[3] in the mouth
2-Methoxy-3-isobutyl-pyrazine	0,000002	–	0,00001
Methyl mercaptan	0,00002		0,00002
Ethyl butyrate	0,001	–	0,0001
Decanal	0,0001	–	0,0009
2-Methylpropanal	0,001	0,01	0,001
Vanillin			0,02
Diacetyl	–	0,007	0,004
Dimethyl sulphide	–	0,001	0,005
2-Phenylethanal	0,004	–	0,009
Hexanal	0,005	–	0,02
Ethyl acetate	–	0,6	0,1
3-Methylbutanol	–	0,8	0,3
Hexanol	0,7	–	0,5
Butyric acid	0,2	–	3
2-Methylpropanol	–	3	8
Ethanol	100	900	10
Pyrazine	300	–	175

[1] D. G. Guadagni et al. 1963; D. G. Guadagni et al. 1972
[2] E. J. Mulders 1973
[3] M. Rothe et al. 1972

components especially. Their low threshold is assumed tq be due to the fact that organic sulphur compounds play a key role in the aroma of many foods (L. Schutte 1974).

In general, acids are less active in aroma efficacy, apart from the sour taste which some of them produce in the mouth. The first member of a homologous series of chemical compounds shows a low aroma efficacy (M. Rothe et al. 1972).

A study of the data summarized in table 19 leads to the conclusion that there is no simple relationship between chemical strućture and aroma efficacy. Extensive studies by R. M. Seifert et al. (1970) on substituted paryzines and by R. G. Buttery et al. (1976) on substituted thiazoles support this observation. From table 20 we can see that insignificant variations in the molecule of pyrazines (for example, the introduction of a methoxy group), or the

Table 20
Threshold concentration of pyrazines and thiazoles in dependence on the substitution (R. M. Seifert et al. 1970, 1972; R. G. Buttery et al. 1976; in ppm)

Substitution of pyrazine	Threshold concentration	Substitution of thiazole	Threshold concentration
none	175	4,5-dimethyl	5
2-methoxy	0,7	4,5-diethyl	0,005
2-methoxy-3-methyl	0,004	4-ethyl-5-propyl	0,00006
2-methoxy-3-ethyl	0,0004	4-ethyl-5-butyl	0,0001
2-methoxy-3-propyl	0,000006	4-butyl-5-butyl	0,0002
2-methoxy-3-isopropyl	0,000002	4-butyl-5-propyl	0,000003
2-methoxy-3-hexyl	0,000001	4-butyl-5-ethyl	0,00002
2-methoxy-3-isobutyl-5-methyl	0,0003		
2-methoxy-3-isobutyl-6-methyl	0,003		
2-methoxy-3-isobutyl-5,6-dimethyl	0,3		

shortening of side-chains in the case of thiazoles, changed the aroma efficacy by orders of magnitude.
Another fact complicates this problem further. Differences in aroma quality and aroma efficacy have been found between some geometric or optical isomers. Menthol, anethol, carvone and nootkatone are examples of such aroma differences between isomers.. In this area research work has recently been published on the gas chromatographic separation of chiral secondary alcohols and their esters, octalactones and decalactones (R. Tressl and K.-G. Engel 1984) and by A. Mosandl and G. Heusinger (1984) with phenylglycidate derivatives.

2.4.3. Weighting problems

The large differences observed between aroma components in aroma efficacy and in threshold value lead to the problem of weighting the importance of these substances in the total flavour impression. This question arose very early in aroma research, and it seems more and more to be a problem of central interest. Though several strategies have been developed to deal with the problem, there is so far no clear answer to the question of which of the numerous volatiles found in aroma complexes are of especially great importance, which of less importance and which others do not participate at all.
Table 21 summarizes some of the approaches to this problem. The oldest test for weighting is the use of **group reagents**. They were applied either directly or in connection with gas chromatographic separation, in the form of reaction gas chromatography (see p. 49). Such reagents, which react rapidly and quantitatively with special classes of aroma components to form odourless derivatives, can be helpful in deciding whether a special class of components is of high or of low importance for total flavour. Depending on the decrease in the typical aroma of a tested food or aroma concentrate, the chemical classes of aroma components can be checked by different reagents. Meat aroma concentrates, for example, show a strong flavour decrease after the addition of mercuric salts, which are known to react rapidly with most sulphur compounds. Similarly, bread aroma concentrates show losses in aroma quality and aroma intensity when carbonyl reagents are added. Problems arise, however, from insufficient specifity of reagents and incompleteness of reaction.

Table 21
Possibilities for weighting aroma components and groups

Principle	Effect	Problems
Use of group reagents for functional groups Reaction GC	Changes in sensory quality indicate important functional groups	Reactions with other functional groups are possible
Splitting of the gas stream after GC separation Sniffing single peaks	Sensory proof of important aroma components and peaks	Key substances are rare. Complex sensations are formed by numerous non-typical effects
Calculation of aroma values (odour units) via concentration and threshold value	High aroma values indicate aroma effective components	Simplification of the system often neglects interactions. Above threshold different substances show different trends
Evaluation of aroma indices via statistical methods	Simple objective methods are to be evaluated and can be used within given limits	High time and research consumption. Wrong evaluation possible because of only statistical correlation

Threshold values are another tool for solving the problem of weighting aroma components. Though the threshold value depends on many factors, such as the type of diluting agent, the sensory procedure or — partially — the test panel, such values have been used successfully in the search for key components. An essential requirement for such work is the use of data obtained under standard conditions. Quite large differences in threshold values have been found by different authors (M. C. MEILGAARD 1974, N. BARYLKO-PIKIELNA et al. 1981).

For evaluating quantitative aroma data in this way, the concept of so-called **aroma values** or **odour units** (M. ROTHE and B. THOMAS 1963, D. G. GUADAGNI et al. 1966, E. J. MULDERS 1973) requires a combination of these threshold data with the concentration of the aroma compound to be tested. Following the concept of S. PATTON and D. A. JOSEPHSON (1957), the quotient between actual concentration and threshold concentration is used.

$$\text{aroma value} = \frac{\text{concentration}}{\text{threshold concentration}}$$

The aroma value resulting from this calculation characterizes the multiples by which threshold concentration of the aroma compound under test is surpassed. The comparison of the aroma values of different aroma compounds enables a ranking of these substances with respect to their importance for total flavour.

Table 22 represents an example from the work of D. G. GUADAGNI et al. (1972) on potatoe chips. Threshold values here were determined in tasteless vegetable oil, as lipids predominate in this snack food.

According to table 22 there are great differences in the contribution of the 18 aroma components tested to the total aroma. Methional (3-methyl-thiopropanal) ranks first in aroma value. According to the concepts previously mentioned, a certain key character of this substance must be supposed. Indeed, among all aroma components studied a sensory test panel trained for this problem found methional to be most similar to the flavour of potatoe chips. The results of table 22 again confirm the fact that the predominating volatile compounds in foods need not be identical with the most important ones for the aroma.

Table 22
Relative percentage and 'odour units' of selected aroma components in steam-volatile potatoe-chip oil
(D. G. GUADAGNI et al. 1972)

Compound	Relative % in oil	Threshold in oil per 10^9 parts	Odour unit $\times 10^3$
Methional	2,0	0,2	100000
2-Phenylethanal	18,0	22	8180
3-Methylbutanal	5,0	13	3850
2-Ethyl-3,6-dimethylpyrazine	7,4	24	2720
2,4-Decadienal	7,5	135	560
2-Methylbutanal	7,4	140	530
2-Ethyl-5-methylpyrazine	6,0	320	190
1-Penten-3-one	0,1	5,5	180
Hexanal	2,1	120	175
2-Methylpropanal	0,5	43	120
2-Nonenal	1,5	150	100
2,5-Diethylpyrazine	1,0	270	37
2,5-Dimethylpyrazine	6,5	2600	25
Heptanal	0,6	250	24
2-Octanal	0,7	500	14
2-Heptenal	1,8	1500	12
2-Decenal	1,2	2100	6
Nonanal	0,1	1000	1

For this type of evaluation it is not necessary, however, to know the structure and identity of every single component. D. G. GUADAGNI et al. (1966) and R. G. BUTTERY et al. (1968) were able to indicate the contribution of the hydrocarbon fraction or of oxygen-containing terpenes to the total aroma of carrot oil and hop oil, respectively.

A classical study in this field is the interesting work of D. G. GUADAGNI et al. (1966) and R. A. FLATH et al. (1967) on the aroma of 'Golden Delicious' apples. Figure 41 uses their data to show the most important gas chromatographic and sensory results obtained with an apple aroma concentrate, summarized in a graphical picture.

Comparing concentration and sensory effect, one can easily state that the very small gas chromatographic peak no. 11 is characterized not only by a typical apple-like aroma but also by a very high aroma efficacy. The 'odour unit' calculated from the percentage amount of the whole aroma concentrate and the threshold value indicate a rather high efficacy compared with the other peaks of the chromatogram.

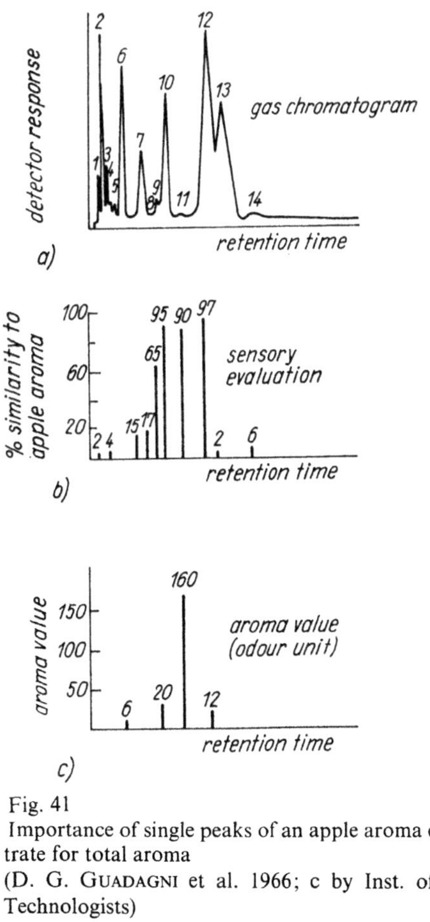

Fig. 41
Importance of single peaks of an apple aroma concentrate for total aroma
(D. G. GUADAGNI et al. 1966; c by Inst. of Food Technologists)

Peak 11 was identified by the authors as ethyl-2-methyl-butyrate. This component obviously has key character in the aroma of the 'Golden Delicious' apple. The usefulness of such calculations is especially clear in this example. The knowledge of a key component in the aroma complex is not restricted to its scientific interest, or its application for flavouring purposes (J. L. BOMBEN et al. 1969): such knowledge can be also used for quality control of technical concentration procedures. The range with the highest similarity to the typical apple aroma in the chromatogram of figure 41 correlates best with the area between peaks 9 and 13. Consequently this area must be suitable as an index for estimation of aroma intensity and quality.

The efficacy of such evaluation methods should not be overestimated, however. Mistakes and wrong conclusions can arise from over-simplified calcularions based on the ceoncept of additivity between the different aroma compounds. Physiological as well as physical factors were in the past sometimes neglected in order to realize a simple calculation.

E. P. KÖSTER and O. WOUTERS (1970) criticize the general supposition of an additive effect between aroma components. If synergistic or antagonistic interactions occur, the results must obviously be distorted. The literature offers some examples of an additive effect (E. A. DAY et al. 1963, D. G. GUADAGNI et al. 1966a, 1966b, R. G. BUTTERY et al. 1968, T. J. SIEK et al. 1969, E. J. MULDERS 1973) as well as of the contrary hypothesis (J. E. LANGLER and E. A. DAY 1964, D. J. KENDALL and A. J. NEILSON 1966, L. L. YOUNG et al. 1970). Substances which are similar in aroma seem to add their aroma effect in most cases, obviously. The likelihood of synergistic and antagonistic effects between aroma components increases with the qualitative difference in aroma, but these mutual influences are valid only in some cases (M. C. MEILGAARD 1975).

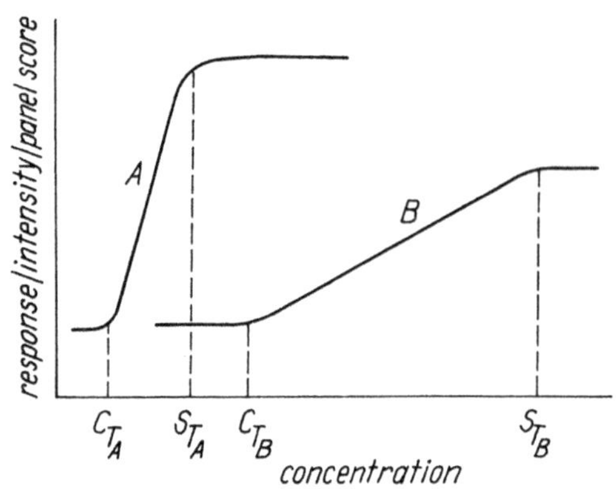

Fig. 42
Idealized relationship between concentration and physiological response for aroma components
(D. J. Casimir and F. B. Whitfield 1978; logarithmic axis)

The concept of aroma values has also been used for the determination of such synergistic or antagonistic effects themselves. P. Salo et al. (1972) and P. Salo (1973, 1975), dealing with a whisky imitation by combining various aroma components, proved significantly that the supplementation is additive in most cases. But there was also evidence that the effectiveness of all components together exceeded the value of the summarized aroma value data. J. Solms similarily found thiamine and thiamine diphosphate producing a synergistic effect in the formation of yeast flavour.

There is another physiological aspect of such calculations which can be criticized. The comparison of the aroma effects of different aroma components starts from the assumption that the aroma increase above the threshold value follows the same function for all aroma compounds. Only in this case would aroma values be comparable. In fact, there are quite large differences in these functions (W. S. Cain 1969, D. G. Land 1979, N. Barylko-Pikielna et al. 1983). The relationship between concentration and aroma effectiveness can be better described by three parameters: threshold value, saturation threshold and the slope of the line between these two points (see figure 42). D. J. Casimir and F. B. Whitfield (1978) define the calculated magnitude of this slope combined with concentration as the 'flavour impact value'. By use of such calculations they found 6-(2-butenylidene)-1,5,5-trimethylcyclohexene, hex-3-enyl butanoate and hexyl butanoate to be the most important aroma compounds of passion fruit.

There are also physical factors influencing the type and intensity of aroma perception in foods. Thus another point for criticism regarding the calculation of aroma values is the use of threshold data which were determined in simple model systems because such models are easy to standardize. Most real food systems are much more complicated. Consequently the threshold data found for

Table 23
Detection limits of aroma components in water and wheaten bread crumb

Aroma compound	Detection threshold (ppm)		ratio water to wheaten bread
	in water	in wheaten bread	
Diacetyl	0.004	10	1:2500
Maltol	0.6	800	1:1333
Ethanol	10	12300	1:1230
Ethylmaltol	0.08	60	1:750
Vanillin	0.02	8	1:400
3-Methyl-butanol	0.7	18	1:26

Table 24
Detection threshold of aroma components in water and vegetable oil
(plastic bottles method; odour-free oil; R. G. Guadagni et al. 1972)

Aroma compound	Threshold concentration (ppm)		ratio water/oil
	in water	in vegetable oil	
Methional	0.2	0.2	1:1
2,5-Dimethyl-pyrazine	1800	2600	1:1.4
2-Phenyl-ethanol	4	22	1:5
Hexanal	4.5	120	1:27
2-Heptanal	13	1500	1:115
Nonanal	1	1000	1:1000
2-Nonenal	0.08	150	1:1875
2-Decenal	0.3	2100	1:7000

Table 25.
Changes in threshold concentration with the fat content of the medium (ppm)

	Water	Skim milk	Milk	Cream
Fat content (%)	0	0.1	2.5	30.0
Vanillin	0.02	0.04	0.04	0.05
Diacetyl	0.004	0.007	0.04	0.07
Furfural	0.04	—	0.4	0.7
Maltol	0.6	0.6	0.8	6

media like water, milk, vegetable oil, beer or wheaten bread differ widely (R. FERON and M. GOVIGNON 1961, D. A. LILLARD et al. 1962, G. A. F. HARRISON and E. COLLINS 1968, P. SALO 1970, M. ROTHE et al. 1972, M. C. MEILGAARD 1975, M. ROTHE 1975, F. A. FAZZALARI 1978).

Interaction has to be considered not only between different aroma components but also with the main nutrients of the food medium. These interactions may possibly influence the vapour pressure over the food in different ways with different intensities. Reasons may be:

- the partial binding effect of aroma components on solid or liquid carriers
- the distribution of the aroma components between an aqueous and a lipid phase.

Tables 23 and 24 support these hypotheses. According to table 24, the aroma efficacy is lower in a fat medium than in aqueous systems; the fat content and distribution coefficient between aqueous and lipid phase must therefore also be of importance (P. E. NELSON and J. E. HOFF 1968, R. G. BUTTERY et al. 1973). Consequently, the threshold concentration in table 25 shows the same ranking order as the fat content: water/skim milk/milk/cream (M. ROTHE 1975).

The practical consequences of this are obvious. Indeed, the flavouring of foods which contain high amounts of lipids requires a much higher amount of aroma concentrates as additives than do low-fat products (A. VAN EIJK 1971). A well-known example is the flavouring of ice-cream. The application level depends on lipid content of the milk base used (J. SCHEIDE 1975).

As well as the lipid/water ratio in foods, the distribution coefficient of aroma components between the water and the lipid phase must influence aroma sensation. Each aroma compound which is in touch with the two phases

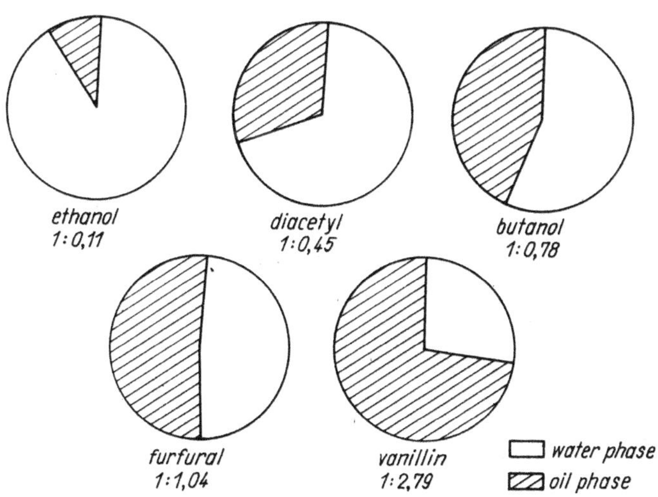

Fig. 43
Distribution coefficients of aroma components between an aqueous and a lipid phase (Distribution between equal parts of water and vegetable oil at 20 °C)

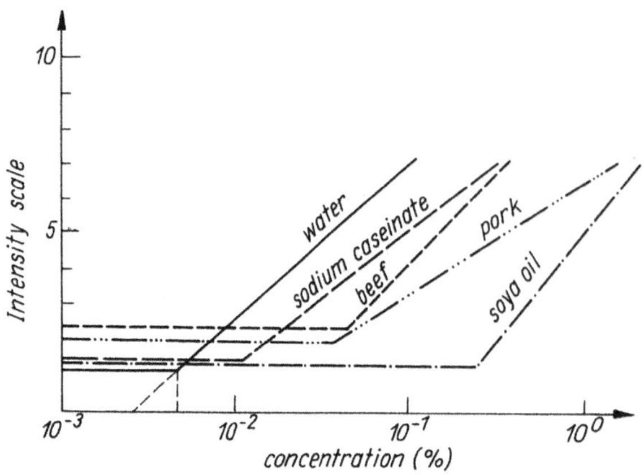

Fig. 44
Relationship between concentration and odour intensity of caraway extract in different media
(I. TYSZKIEWICZ and A. JACKOWSKA 1982)

tends to distribute into both, according to its structure and solubility. With respect to structure and polarity there are, however, large differences between the numerous aroma components (figure 43).

Figure 44 shows the relationship between concentration and odour intensity of a caraway extract in different media (I. TYSZKIEWICZ and A. JACKOWSKA 1982). Large differences in the threshold value were found. There are also differences in the slope of the calculated regression lines which characterize the relationship between concentration and aroma response. Different interactions between concentration and aroma response. Different interactions between the molecules of the medium and those of the aroma components may be the reason.

Recently F. DRAWERT and N. CHRISTOPH (1984) combined the gas chromatographic sniffing technique (see figure 33 and 34) with odour recognition threshold values of the aroma compounds separated. By this procedure limonene, myrcene and octanal were found to be the main participants in the aroma complex of lemon. By a similar concept D. G. CUNNINGHAM et al. (1986) combined odour intensities revealed by sniffing GC effluents with retention data ('charm analysis').

Considering all the factors which can change the threshold values and aroma efficacy, the concept of aroma values may meet with some criticism (E. P. KÖSTER 1975, M. ROTHE 1975). Different types of interaction have to be calculated. Nevertheless, this method provides one of the few possible ways of weighting the numerous aroma components within an aroma complex.

2.5. Relationship between sensory and instrumental data

Following the definition in chapter **1.3.**, that the aroma of foods means the complex aroma impression produced by the sum of volatile aroma components, aroma research should primarily be defined as a physiological or psychophysical problem (E. VON SYDOW 1971). But in fact flavour and aroma research in the past have been mainly in the chemist's domain. To elucidate the problems, modern instrumental techniques were readily available for the necessary separation and identification of all substances composing food volatiles.

Thus, in the years following the introduction of chromatographic and spectrometric methods into aroma research, qualitative aroma analysis took the first place. This was obviously a consequence of the supposition that quantitative data could not be of any interest until the qualitative elucidation of the aroma complex was completed. With the numbers of aroma components continually increasing there seems to be no chance of reaching this point in the near future.

Problems of aroma formation during food processing or of food flavouring, however, need a quantitative evaluation. Small differences in aroma intensity or aroma quality must be quantitatively determined with suitably high reproducibility and significance. This can be realized by sensory analysis, but in principle, instrumental parameters seem to be more suitable for determining such attributes as aroma intensity, aroma concentration, or changes of aroma.

This discussion does not affect the central importance of sensory analysis in flavour research. Only by sensory means can we decide whether a special flavour impression is registered as a positive or negative sensation, as too weak or too strong, as being in accordance with consumer acceptability or not. Thus sensory analysis will remain the most important procedure in aroma research, as well as the only measure for all instrumental data.

When we consider these points, the way to solve most of the practical aroma problems seems to be clear. As a first step, a defined sensory analysis has to be carried out: this must be sensitive and reproducible enough to have high statistical significance. Secondly, the result can be complemented or replaced by instrumental data. Naturally the determination of such intrumental parameters should require as little time and laboratory work as possible. With respect to the complexity of food aromas and the trace concentration of most volatiles, however, this task seems to be difficult enough.

Generally all analytical techniques are suitable for the search for parameters which supplement or replace sensory characteristics. There are many examples in classical aroma analysis. One of the oldest parameters for determination of an off-flavour in foods, for example, is the peroxide value (C. H. LEA 1946, K. TÄUFEL and R. VOGEL 1955, C. FRANZKE 1956). If this value increases in fats or lipid-rich foods and exceeds a certain limit, the product under test is spoiled by oxidative rancidity. The peroxide number, however, is a typical index because the rancid taste is produced not by the peroxides themselves but by their degradation products. These 3—9 carbon alkanals, alkenals and alkanones are normally present in a much higher concentration than the unstable peroxides.

In cases where single components with key character (*character impact compounds*) are known, their concentration will be appropriate as such an aroma index. But, besides the fact that key components are rare in the flavour area, there are also separation and determination problems because key components are often present only in trace amounts. Other types of indices are therefore also of interest. E. VON SYDOW (1975) discussed three groups of such indices:

— **accidental** (*ad hoc*) **indices** which are valid within one single experiment only and thus are of limited value

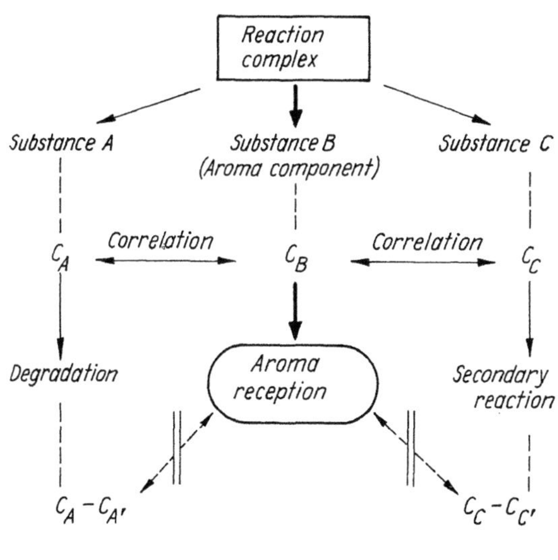

Fig. 45
Aroma index — scheme about effect and limits

Table 26
Aroma components with key or index character found by the 'aroma value' concept

Food	Important aroma components	Reference
Beer	myrcene	D. G. GUADAGNI et al. 1966 a
Apple 'Golden Delicious'	ethyl 2-methylbutyrate	D. G. GUADAGNI et al. 1966 b
Carot	2-nonenal	R. G. BUTTERY et al. 1968
Potatoe chips	3-methylthiopropanal	D. G. GUADAGNI et al. 1972
Whisky	diacetyl	P. SALO et al. 1972
	butanol	P. SALO et al. 1973, 1975
	isopentanal	
	isoamyl acetate	
	ethyl hexanoate	
	ethyl octaoate	
Rye bread crust	2-methylpropanal	M. ROTHE et al. 1972
Wheaten bread	acetaldehyde	E. J. MULDERS 1973
Lemon	limonene	F. DRAWERT and N. CHRISTOPH 1984
	myrcene	
	octanal	
Grapefruit	ethyl butyrate	J. PINO et al. 1985
	limonene	
	decanal	
	nootkatone	

- **predictive indices** which are valid for all samples defined by certain independent parameters
- **causative indices** which act as real stimulants resulting in particular flavour attributes.

Finding causative indices is one of the ultimate aims of the flavour scientist. This task, however, is a difficult one in most cases because it requires a complete knowledge of the chemical and psychophysical basis.
The principle of using predictive indices as an analytical tool is shown in figure 45. In complex reaction systems, different reaction pathways run in parallel and simultaneously. Supposing that one of these reactions (for example B in figure 45) leads to an aroma component or an aroma mixture which cannot be determined directly without problems, between all three reactions A, B and C, A or C it must also be possible to use the reaction A or C for evaluation. If a correlation is found is appropriate as a predictive index for reaction B or the concentration of B, respectively.
The limits of such indices, however, will be clear: if the three reactions or reaction products behave differently concerning, for example, volatilization, degradation, or any type of secondary reaction, the index concept cannot work exactly. Thus before such predictive indices can be used the limits of application have to be carefully tested and proved.
In table 27 some examples of aroma indices used or proposed in the literature are given. There are direct and indirect related indices. Such parameters as linalool for the muscat-like aroma of special dessert wines, or 2-heptanone for the Roquefort note of cheeses, may be causative and thus directly related. Typical of an indirect relationship is pyruvic acid, which itself lacks aroma, as an index for the sharpness of onion powders. From figure 46 we see that the acid is produced during cell degradation from S-alkyl-cysteine sulphoxides formed in parallel with the production of the sharp onion-like and lachrymatory attributes (S. SCHWIMMER et al. 1964, R. TRESSL 1975).
Negative aroma sensations, such as off-flavour or taints, can be discussed in the same way. Such indices are in use for some types of food spoilage and shelf life: examples are shown in table 28.
But apart from the good separation effects of the aroma index concept, a significant quantification of data is necessary. Small quantitative differences have to be recognized

Table 27
Indices characterizing food flavour and aroma or their attributes

Food	Aroma index	Characterized attribute	Authors
Citrus juices	H_2S	freshness	R. E. BERRY et al. 1983
Tomatoe juice	dimethyl sulphide	aroma intensity	D. G. GUADAGNI and L. C. MIERS 1969
Onions/onion powder	pyruvic acid	onion sharpness	S. SCHWIMMER and W. J. WESTON 1961
Beer	myrcene	hop bitter taste	D. G. GUADAGNI et al. 1966; D. R. MAULE 1967
Cider	isoamyl acetate	cider aroma	K. WUCHERPFENNIG and G. BRETTHAUER 1968
White wine	methyl hexanoate	Riesling aroma	A. C. NOBLE 1976
Dessert wine	linalool	Muscatel aroma	L. USSEGLIO-TOMASSET 1969
Brandy, whisky	C_4 and C_5 alkanols	aroma quality	K. KAYAHARA et al. 1966; J. A. ZEE et al. 1984
Wheaten bread	C_4 and C_5 alkanols	fermentation aroma	G. WÖLM et al. 1974
Pumpernickel	furfural	aroma intensity	M. ROTHE ET AL. 1967
Blue cheese	2-heptanone	Roquefort intensity	M. ROTHE et al. 1981

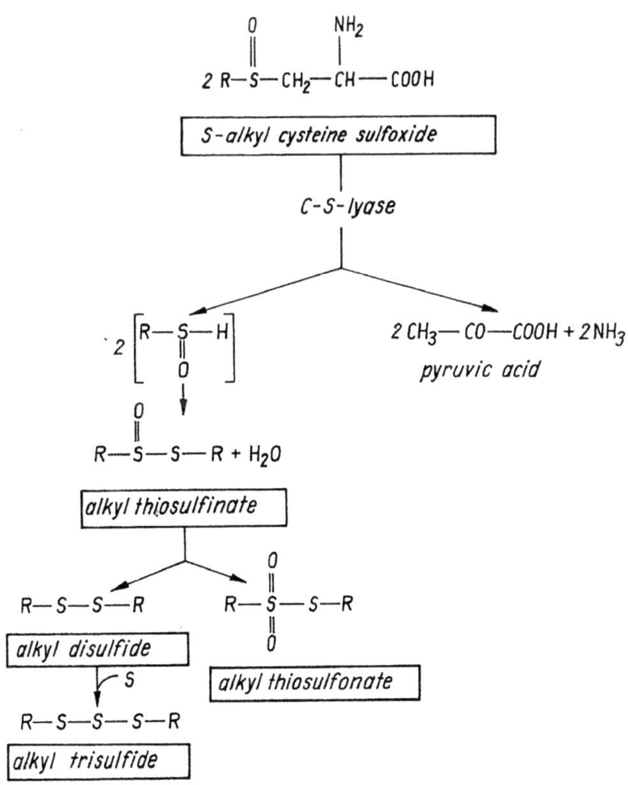

Fig. 46
Enzymatic degradation of S-alkyl sulphoxides to sulphur-containing aroma compounds and pyruvic acid (A. STOLL and E. SEEBECK 1951; R. TRESSL et al. 1975)

by the methods used (J. SCHÄFER and C. WEURMAN 1970). Furthermore, all methods for the determination of aroma indices should be easily and rapidly performed, because clear conclusions often require a large amount of data. When we further consider the high number of peaks in a gas chromatogram, it becomes obvious that statistical methods and computer analysis are required for optimum work in this field.

The recent literature in flavour research indicates several examples of systematic experiments designed to find appropriate aroma indices. Some of these publications are summarized in table 29. It seems to be clear from this work that most progress is made when the relationship between sensory and instrumental data is studied in detail. On the other hand, the development of suitable methods usually needs a lot of time and research capacity, on account of the large data pool required. Computer techniques sometimes have to be used.

A lot of research work in the last 15 years has dealt with the relationship between sensory and instrumental data. Because of the different methods of statistical analysis used, it is not possible to review these interesting activities in detail. A summary of a few typical publications is given in table 29.

Some examples, however, will be discussed. One of the first groups to enter this field of systematic research was that of E. VON SYDOW in Göteborg in the early 1970s. One topic was the analytical proof of aroma changes occurring in canned meat products during processing or storage (T. PERSSON et al. 1973, E. VON SYDOW 1975). As a first step in this ambitious study, two parallel tasks were carried out separately: the gas chromatographic and mass spectrometric analysis of the whole system, and a quantified sensory descriptive analysis. From 95 gas chromatographic peaks originally separated and identified, 22 were included in the evaluation work. A sensory panel after training was able to differentiate and quantify 15 sensory attributes in canned meat.

Correlation analysis and model functions were used to find relationship between the two groups of data. According to the functions of WEBER-FECHNER and STEVENS, the relationship between stimulus concentration

Table 28
Indices characterizing off-flavour in foods

Food	Characterized off-flavour	Aroma index	Authors
Cereal grain	mouldy or musty odour	fat acidity lipase activity	D. BAKER et al. 1959; M. ROTHE et al. 1967
Wheat germs	bitter taste	fat acidity	M. ROTHE and J. STÖCKEL 1961, 1962
Fats and oils	rancidity	peroxide value	K. TÄUFEL and R. VOGEL 1955; C. FRANZKE 1956
Baked and instantized cereal products	rancidity	medium chain fatty aldehydes	M. B. BOGGS et al. 1966; M. ROTHE et al. 1967
Wheaten bread	ropy odour *(Bacillus subtilis)*	diacetyl	M. ROTHE 1963
Sauerkraut	cheese-like off-flavour	propionic, butyric and caproic acid	M. L. VORBECK et al. 1961
Apple juice	off-flavour	ethanol	J. KOCH 1972
Apple juice, orange juice,	aroma decrease cooked off-flavour	furfural 5'-hydroxymethyl-furfural	J. HERRMANN 1971; H. E. NURSTEN and M. L. Woolfe 1972; S. NAGY and H. L. DINSMORE 1974
Grapefruit juice	oxydation flavour	α-terpineol	J. PINO et al. 1985
Milk	'sunlight' off-flavour	methional	S. PATTON 1954
Canned meat	canned off-flavour	H_2S, methyl mercaptan, dimethyl sulphide	E. VON SYDOW 1975

(S) and stimulus response (R) can be of a semilogarithmic or exponential nature, in addition a linear model was included:

$R = A \cdot S + B$ (linear model)

$R = A \cdot \log S + B$ (model according to WEBER-FECHNER)

$\log R = A \cdot \log S + B$ (model according to STEVENS)

Within the complex sensation observed during consumption of meat products, with more than 500 identified aroma compounds, it must be supposed that stimuli are produced not only by single substances but also by interrelation and combination effects between components. Thus simple mathematical ratios and functions were also included in the mathematical evaluation program. Some of the results with linear psychophysical models are shown in table 30.

The correlation coefficients in table 30 lead to the conclusion that some sulphur components, such as hydrogen sulphide, methyl mercaptan, dimethyl sulphide and thiophenes, as well as some simple alkanals, are correlated with the off-flavour of canned meat and thus can be used as an index. The authors could prove that this index is not only predictive but to some extent causative, by the addition of substances which are known to bind sulphur components or aldehydes. Reducing the volatiles in question by these reactants also decreased the off-flavour. Computer

Table 29
Selected literature dealing with use of statistical methods for the elucidation of relationships between gas chromatographic and sensory data

Method	Problem	Food	Authors
Simple regression and correlation	Aroma quality and intensity	tomato and potato products	D. G. GUADAGNI and J. C. MIERS 1969
	GC peaks and quality	beer	A. D. HAUKELI et al. 1973
Multiple regression and correlation	Aroma quality and intensity	tobacco	S. SUGAWARA et al. 1971
	Aroma quality and intensity	canned beef	T. PERSSON et al. 1973
	Aroma quality and intensity	black currants	G. KARLSSON-EKSTRÖM and E. VON SYDOW 1973
	Variety differences	ginger oil	A. A. BEDNARCZYK and A. KRAMER 1971
	Aroma evaluation	hop	E. KRÜGER and L. NEUMANN 1975
	Quality differences	orange oil	G. F. GOSTECNIK and A. ZLATKIS 1975
	Aroma evaluation	apple juice	R. SCHRÖDTER 1976
Discriminant analysis	Flavour differences, classes	roasted coffee	J. J. POWERS and E. S. KEITH 1968
	Variety and quality classes	coffee	R. E. BIGGERS et al. 1969
	Quality changes	tomato products, nuts	L. MILUTINOVIC et al. 1970
	Quality evaluation	cola beverage	L. L. YOUNG et al. 1970
	Variety influence	peanuts	H. PATTEE and J. SINGLETON 1972
	Storage influence	olive oil	J. M. O. JIMENEZ et al. 1974
Factor analysis	Quality characterization	tea	L. VUATAZ and D. REYMOND 1971
	Quality parameter	beer	L. REINER and A. PIENDL 1973

Table 30
Correlation coefficients between sensory attributes and gas chromatographic peaks connected with canned meat off-flavour
(E. VON SYDOW 1975; linear model; single peak and 2 peak combinations)

	Odour qualities			
	Nasty-smelling	Cooked vegetables	'Retort flavour'	Sulphurous
Peak combination: x		x: 2-methylbutanal r = 0.57	x: methyl mercaptan r = 0.70	
Peak combination: x · y	x: pentanal y: 2.3-butandione r = 0.75	x: dimethyl sulphide y: thiophene r = 0.81	x: H_2S y: 2-methyltiophene r = 0.80	x: methyl mercaptan y: 2-methylthiophene r = 0.82
Peak combination: x/y	x: 3-methylbutanal y: dimethyl sulphide r = 0.68	x: 3-methylbutanal y: pentanal r = 0.83	x: 2-methylbutanal y: ethylene sulphide r = 0.82	x: methyl mercaptan y: heptanal r = 0.84

analysis was necessary because of the large volume of data obtained and calculated.
A similar concept was used in another project of the same Göteborg group, dealing with the aroma of black currants juice (E. VON SYDOW et al. 1970, E. VON SYDOW and G. KARLSSON 1971, G. KARLSSON-EKSTRÖM and E. VON SYDOW 1973; see also figure 33). Table 31 includes those correlation coefficients which were found to be above 0.70. Obviously there is a highly positive or negative correlation between several single peak areas and special attributes of sensory quality. The concentration (peak area) in phellandrene and cis-β-ocimene, for example, is negatively correlated with such properties as 'cooked vegetables' and 'musty'. The same components gave a positive correlation to notes like 'fruity', 'aromatic' or 'fragrant'. Some negative factors of sensory quality like 'sick', 'fermented', 'sharp', 'acid' or 'boiled flavour' seem to depend directly on the concentration of

Table 31
Correlation coefficients between selected sensory and instrumental data found in connection with heating of black currants
(G. KARLSSON-EKSTRÖM and E. VON SYDOW 1973; function according to STEVENS' Law)

	Pleasant sensory notes		Unpleasant sensory notes	
	fruity	fragrant	musty	cooked vegetables
Compounds increasing with heating:				
dimethyl sulphide	−0.86[1)]	−0.85	0.86	0.88
2-methylpropanal	−0.86	—	0.85	0.86
α-terpineol	−0.87	−0.86	0.82	
Compounds decreasing with heating:				
β-phellandrene	0.87	0.86	−0.86	−0.87[1)]
caryophyllene	—	0.82	−0.82	—
cis-β-ocimene	0.81	0.85[1)]	−0.86[1)]	−0.86[1)]
terpinolene	—	0.83	−0.84	−0.87[1)]

[1)] = function according to FECHNER's Law

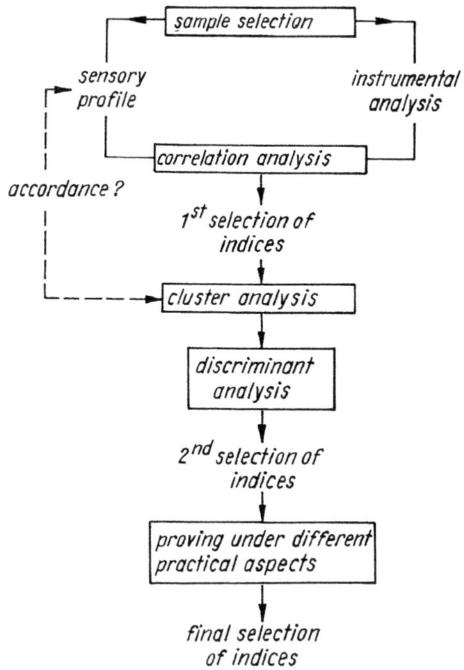

Fig. 47
Scheme of selecting aroma indices appropriate for flavour quality problems
(M. ROTHE et al. 1982)

methyl sulphide or of some simple alkanals.

As a result of a very intensive and wide-ranging research program, aroma indices result which are relatively easily to determine and which are of practical value: flavour changes in blackcurrants caused by pasteurizing, for example, can be proved by determination of only a few components. In this respect the decrease in special terpene hydrocarbons seems to be as typical as the increase in dimethyl disulphide or in simple alkanals.

Another example comes from our own research work on Blue cheese flavour. The scheme shown in figure 47 was used (M. ROTHE et al. 1982). As a first step, 20 samples of Blue cheese were selected in such a way that the differences in sensory quality were related to the question to be answered. In practice the products of one factory were used, taken from different production batches and differing in ripening time.

In the main experimental step all samples were analysed both by sensory profile analysis and by instrumental methods including gas chromatography of fatty acids, head space analysis of the main gas chromatographic peaks, methyl ketone analysis by thin layer chromatography, and some classical values. Correlation analysis was the first step in statistical evaluation. Table 32 shows an abstract of high correlation coefficients between the two groups of data. From these results, those instrumental values which were highly correlated were selected for a cluster analysis. We had to answer the question whether this type of statistical analysis would yield the same quality and intensity data as direct sensory evaluation. The calculation of the quality groups was based on 20 selected analytical values.

One of the dendrograms obtained as a result of the clustering is shown in figure 48. It indicates four quality groups with 7, 9, 3 and 1 samples, respectively. The intensities of the 'rancid' note (one of the most important sensory attributes of Roquefort-type cheeses) shown in the same figure demonstrate exactly the same quality groups. This was found to be true also for the 'Roquefort' note and for the 'soapy' note of the cheeses, with a maximum of two wrong groupings out of a total of 20.

We can conclude from these that the instrumental values selected and used for the cluster analysis may be appropriate as an aroma index. This opinion can be supported by the calculation of average values and standard deviation from the samples within the clusters obtained. Table 32 shows how this is done: when the calculation results in a higher standard deviation with overlapping between the clusters, the index tested is not very appropriate. However, two of the values included within the table show a rather good separation effect. They can be taken as an aroma index within the limits given by the selected samples and by the analysis program (see also M. ROTHE et al. 1981).

Discriminant analysis and factor analysis are further statistical methods used for research about indices, being appropriate for instrumental quality analysis. There are many examples in the literature of the last 15 years (see table 29).

The most important result of such research is not the knowledge about relationships in

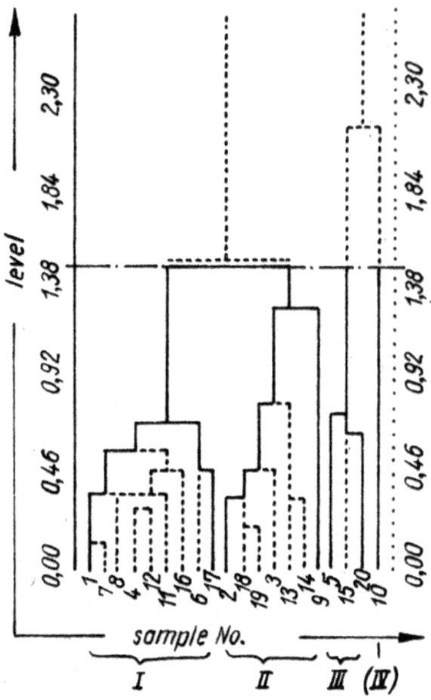

Cluster No.	rancid intensity
I	0,7 0,8 1,3 1,5 1,6 2,1 2,1
II	2,2 2,2 2,4 2,5 2,6 2,6 2,6 2,8 3,1
III	3,7 3,7 4,4
IV	5,0

Fig. 48
Classification of instrumental data obtained on 20 samples of Blue cheese by cluster analysis and comparison with sensory grouping
(M. ROTHE et al. 1982)

Table 32
Appropriate instrumental indices for characterization of the 'Roquefort' intensity determined by cluster analysis
(M. ROTHE et al. 1982)

Instrumental index	Cluster average ± s			Suitable for differentiation[1]
	Cluster I 'strong'	Cluster II 'medium'	Cluster III 'slight'	
Acetoin	1903 ± 167	2305 ± 738	542 ± 326	(+)
Headspace peak 4	562 ± 227	478 ± 249	168 ± 87	—
Headspace peak 10	61 ± 13	81 ± 62	9 ± 5	(+)
Mono-carbonyls	0,91 ± 0,23	0,51 ± 0,1	0,42 ± 0,1	(+)
Methyl ketones	0,70 ± 0,18	0,25 ± 0,03	0,17 ± 0,05	+
Acetone	15 ± 5	10 ± 2	5 ± 3	—
2-Pentanone	101 ± 38	32 ± 13	13 ± 6	+
2-Heptanone	166 ± 52	51 ± 20	42 ± 31	(+)
2-Nonanone	141 ± 46	48 ± 14	32 ± 23	(+)
		0,27 ± 0,08	0,25 ± 0,06	(+)
2-Heptanone/2-pentanone	1,6 ± 0,2	1,6 ± 0,2	3,7 ± 3,8	
Yellow index[2]	5,2 ± 1,1	3,8 ± 1,9	2,0 ± 0,7	

[1] Differentiation: + = appropriate, (+) = possible, — = impossible
[2] Sum of volatile carbonyls

the field of flavour and aroma analysis itself. Significant indices for flavour quality or its attributes, however, may be a very useful tool for all the practical problems of aroma production, aroma and off-flavour control, aroma retention and degradation. After a long period of identification work in the complex field of food flavours, which sometimes looked like degenerating into a mere catalogue of numerous substances, effects and interactions, we may now have a chance to simplify some of the problems involved. Thus we can assume that aroma indices will have a prominent place in modern aroma research.

3. General problems in aroma research

3.1. Chemistry and biochemistry of aroma formation

The complexity and inhomogenity of food aromas (see table 18, p. 58) lead to the assumption that the single aroma components must be formed by very different and specialized reaction mechanisms. The distribution into chemical classes provides little more information than defining the most important chemical groups or — as an exceptional case — the key classes. Comparing the lists of volatile compounds in food collected and summarized by S. VAN STRATEN et al. (1973, 1977), H. MAARSE (1984, 1985) and H. MAARSE and C. A. VISSCHER (1986), however, some similarities can be found. It may be surpising, for example, that some very simple aroma components have been identified in practically all foods. This is true for all saturated alcohols, aldehydes, and amines with up to six carbon atoms and the even-numbered fatty acids of short to medium chain length, as well as the first members of homologous series, for example acetone, 2.butanone, 2,3-butanedione (diacetyl), methyl formate, methyl and ethyl acetate or methyl mercaptan. Additionally, a lot of other components are also found in very different food items. The total number of aroma compounds identified is limited by such effects: nearly 4500 volatiles have been identified in food up to now, but this is a relatively small number compared with the fact that in every food item examined in detail between 100 and 600 components have been found.

Considering this wide spread of aroma compounds within the broad field of foods and beverages, with their very specialized flavours, the question arises whether a typical aroma might be a consequence of different amounts of the same substances. Cocoa was earlier discussed as an example. Of the 310 aroma components identified by 1976, 260 were also found in other foods (W. MOHR et al. 1976). Might this fact be caused by the same, or similar, formation mechanisms?

Some other specific details from the very informative material compiled by S. VAN STRATEN et al. (1973, 1977) and H. MAARSE (1984, 1985) also desire a mention here. A broad spectrum of terpenes is well-known and characteristic of most of the essential oils and spices. However, recently such compounds have also been found to be of key character for some fruits and vegetables. Blackcurrants, bilberries, carrot and celery are examples. A very high amount of volatile sulphur components seems to be typical not only for onion and cabbage but also for heated meat products or roasted coffee. Volatile esters, which are typical ingredients of fruits, have also been found in foods and beverages where fermentation plays a dominant role in processing: cheese, cocoa and bread are examples.

For all food products where the production process includes heat treatment, a high number of basic components can be found in the spectrum of identified volatiles (coffee, cocoa, potato chips, roasted peanuts, bread, cooked or roasted meat). When heated foods contain fat in high amounts, series of even-numbered

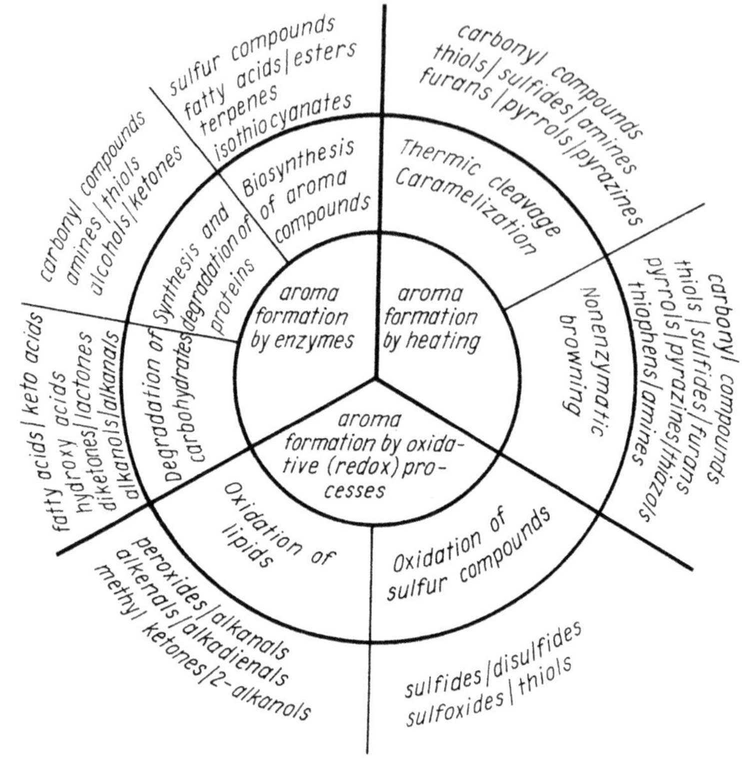

Fig. 49
Most important reaction types of aroma formation

2-alkenals and 2,4-alkadienals with a chain length of 3 to 12 as well as methyl ketones between C_3 and C_9 have been found.

Should we therefore expect joint formation mechanisms? Can we find here a clue to setting the very heterogeneous aroma spectra in order?

It is difficult to give a clear answer, but if some simplifications are allowed the formation principles of aroma compounds can be classified according to the scheme of figure 49 (M. ROTHE 1974). This arrangement divides the field of aroma formation into three parts, i. e. enzymatic, thermic and oxidative reaction mechanisms.

In many foods of plant origin, aroma components are formed during the growing and ripening period. Normally this pathway consists of a rather complicated **enzymatic aroma formation** based on the biosynthesis or degradation of nutrients like carbohydrates or proteins. Aroma-forming mechanisms, however, are side-chain reactions in the micro range which normally run parallel to a lot of other different and complex reaction sequenses. Because of the trace character and complexity of the aroma forming mechanisms, it has been rather difficult to elucidate the origin of food aromas. The chances of acquiring more knowledge in this field have been greatly improved by the modern instrumental techniques of separation and identification. Conversely, the aroma efficacy of most aroma components has been a great help in finding the right way in the very complex labyrinth of chemical reactions and substances.

In the large field of enzymatic aroma formation, secondary reactions are of special importance in plant metabolism. Such effects are often involved when a mechanical or thermic interruption of cell structures occurs during food processing or the preparation and cooking of meals. In the case of onion or garlic, for example, aroma production is directly related to this cell degradation. The formation of volatile sulphur components with intensive flavour and lachrymatory attributes starts at the moment of cutting or homogenizing (see figure 46 and table 27;

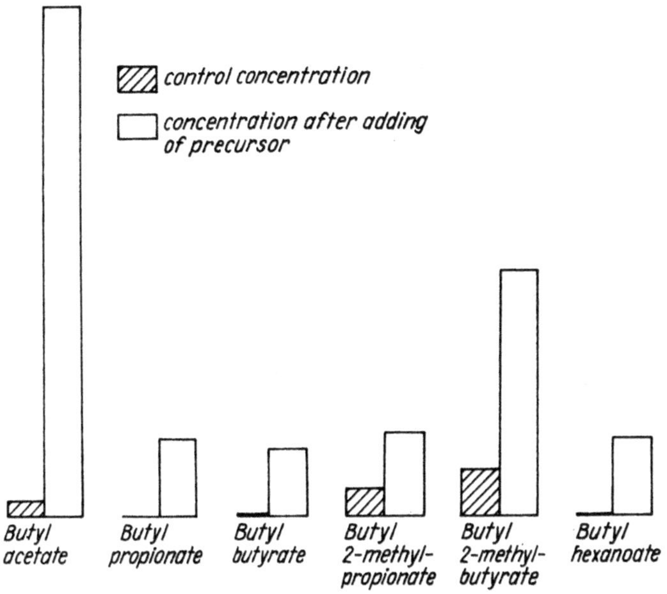

Fig. 50
Formation of butyl esters in strawberry fruit slices after addition of 2-ketovaleric acid (Example from work of F. DRAWERT and R. G. BERGER 1982)

S. SCHWIMMER and D. G. GUADAGNI 1962, A. I. VIRTANEN 1965, L. SCHUTTE 1974, R. TRESSL et al. 1975).

In the case of fruits, aroma formation as well as aroma degradation can be caused by mechanical cell discruption. The large group of esters is probably synthesized in fruits in connection with the β-oxidation of fatty acids. Alcohols formed by a reductase reaction from the corresponding aldehydes react with acyl-CoA to give the ester and CoA. R. TRESSL and F. DRAWERT (1973), F. DRAWERT and H. J. KÜNANZ (1975), I. YAMASHITA et al. (1977) and F. DRAWERT and R. G. BERGER (1981), (1982a, b) proved this hypothesis by addition of precursor substances and use of optimum incubation parameters with whole fruits and with slices of bananas, strawberries and citrus fruits. As exogenous precursors L-alanine, acetate, pyruvate, n-butyrate, 2-ketovalerianate, n-butanal, n-hexanoate and D,L-mevalonic acid-γ-lactone were successfully used. As an example, figure 50 shows the effect of added 2-ketovaleric acid on the formation of butyl esters in incubated strawberry slices. Capillary gas chromatographic methods have been used for separation.

In contrast to such biosynthetic effects, a typical aroma degradation is known for mashes produced from fresh fruits. Here the aroma intensity decreases in dependence on storage time and temperature; this is caused by the activity of hydrolytic enzymes. Frozen fruits show only small degradation effects immediately after thawing (P. SCHREIER 1980), but the aroma biosynthesis is inhibited (F. DRAWERT and R. G. BERGER 1983).

Because of the isoprene-based structure of most volatiles in essential oils, this field seems at first sight to be somewhat less complicated. In all such products we find, however, a highly differentiated system of terpenes and sesquiterpenes. Some interesting recent publications have started to elucidate the 'black box' of complicated biochemical reactions. P. J. WILLIAMS et al. (1980, 1982) with muscat grapes and K.-H. ENGEL and T. TRESSL (1983) with passion fruit demonstrated that there are non-volatile precursors of a glycosidic nature which can produce many different terpene voltailes during ripening and processing. K.-H. ENGEL and R. TRESSL found 16 terpene hydrocarbons, alcohols and oxides to be formed from isolated glycosidic precursors. Figure 51 shows the nine main components identified, with linalool dominating by far.

The primary aroma compounds originally present in the raw materials and produced by enzymatic mechanisms are often not very stable under processing conditions. Especially

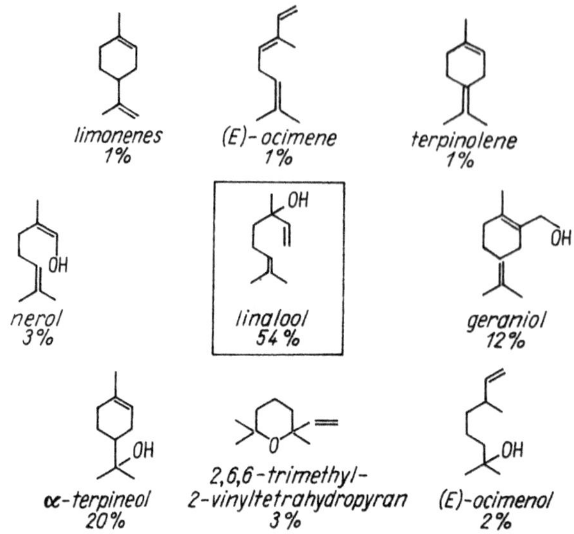

Fig. 51
Volatile monoterpene components formed from isolated glycosidic precursors of passionfruit by enzymatic hydrolysis and heat treatment
(K.-H. ENGEL and R. TRESSL 1983)

in the case of thermic treatment of fruits and vegetables, aroma losses have long been known to occur. Drying processes are a typical example. All earlier technologies for processing of fruits and vegetables reduced the original aroma in both intensity and quality. This effect was a consequence of volatilization during the intensive drying procedures which were applied to reduce the naturally occurring high amounts of water in an economic way.

An interesting idea for aroma regeneration in such cases was published as early as 1956 (E. J. HEWITT et al. 1956, W. RÖDEL 1971, H. GREMLI and J. WILD 1974). The principle is shown in figure 52. Fresh enzyme extracts from the same plant material are necessary. They react with precursors which are nonvolatile and thus remain unchanged during the drying process. The aroma regeneration has been proved by sensory analysis in laboratory tests, but up to now this process

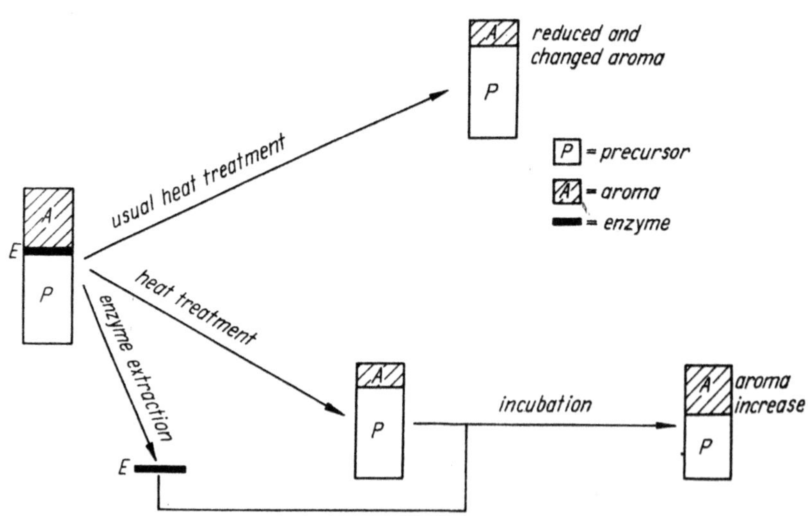

Fig. 52
Principle of enzymatic aroma regeneration following aroma losses caused by drying processes in vegetables
(E. J. HEWITT et al. 1956; column height for intensity)

has not been used in practice. High costs of producing active and specific enzyme preparations as well as lack of basic knowledge may be the reason.

Aroma formation by enzymes is not found only in fruits and vegetables; it is also one of the main aroma-producing principles during food processing. Enzymatically directed fermentation processes are involved in the production of bread and milk products, of alcoholic beverages, as well as of cocoa or of some sausages. During such fermentation, aroma components of high aroma efficacy can be formed as by-products of microbial metabolism. Short-chain or medium-chain fatty acids, diacetyl, higher alcohols, esters or methyl ketones can be formed in this way (compare, for example, R. C. LINDSAY 1967, M. ROTHE 1974, J. E. KINSELLA and D. H. HWANG 1976, L. NYKÄNEN and H. SUOMALAINEN 1983). As an example, figure 53 shows the mechanism which leads to the formation of methyl ketones. The decarboxylation of the β-keto acid is a side-path of the β-oxidation of fatty acids; it may be understood as a detoxification process for the microorganisms, which are sensitive to excess concentrations of free fatty acids.

Another example for the practical utilisation of enzymatic activities in the flavour field is the so-called 'bioconversion' of aroma compounds. By means of bacteria terpenes have been rearranged into such derivatives which are characterized by higher aroma intensity and/or improved aroma quality. The change of bicyclic or acyclic terpenes into monocyclic ones belong to this field, as well as the specific conversion of racemic mixtures or the conversion of valence into the strongly bitter principle nootkatone (compare review of A. LATRASSE et al. 1985).

Another important category of aroma-producing food technologies are reactions with a **thermic aroma formation**. Heat processes are widely used in food production. As the most important mechanism involved, the non-enzymatic browning reaction must be mentioned. It is well-known as the MAILLARD reaction, after the French scientist who discovered and examined this type of reaction as early as 1912. Browning reactions occur during such production steps as cooking, frying, roasting, baking or drying. They are characterized by the brown colour of polymer complexes, defined as melanoidines, which are formed from free amino acids or simple peptides on one hand and sugars on the other. Figure 54 shows that many different substances can be formed in the course of this complicated reaction, with carbonyl compounds as well as with nitrogen, sulphur or oxygen-containing heterocyclic compounds as typical and aroma-effective volatiles. Baking is used as an example; most of the aroma components of bread seem to be products of the non-enzymatic browning reaction. The MAILLARD

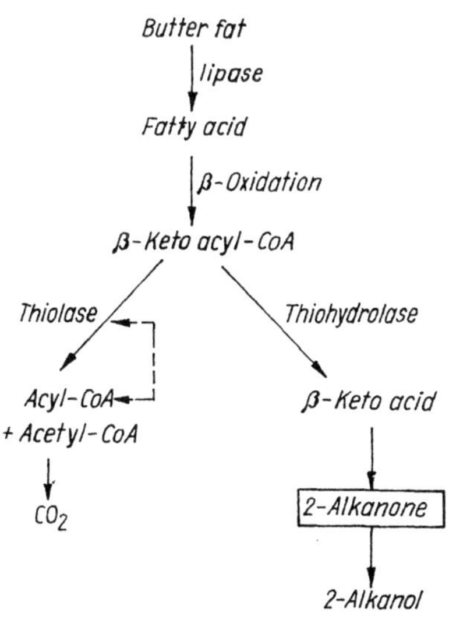

Fig. 53
Formation pathways of methyl ketones by *Penicillium roqueforti*

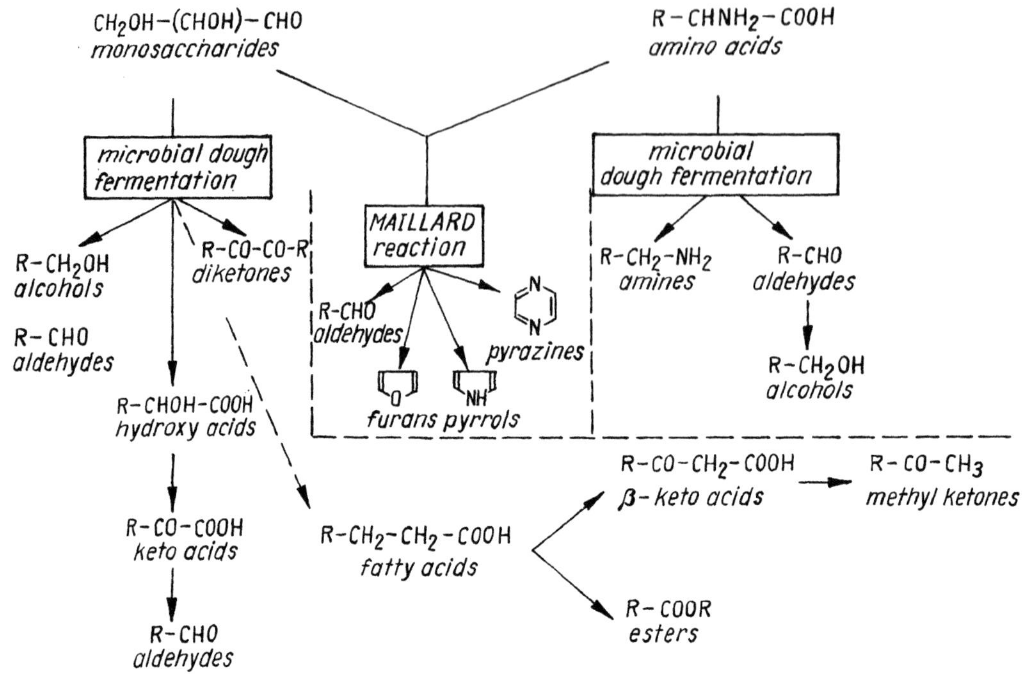

Fig. 54
Principle ways of aroma formation in bread

reaction is discussed in detail in reviews and articles by G. P. ELLIS (1959), J. E. HODGE (1967), J. A. MAGA und C. E. SIZER (1973), R. TRESSL et al. (1974).
Caramelization reactions on the base of sugars also occur during baking. In contrast to the non-enzymatic browning, which can also occur under a lower temperature regime, caramelization needs high heat. In bread, temperatures above 100 °C are reached in the outer layers of the crust only. The mechanism of caramelization seems to be as complicated as that of the MAILLARD reaction, depending however, on the temperature. By caramelizing pure glucose without the presence of other reactants, K. HEYNS et al. (1966) identified 71 volatiles including 29 heterocyclic compounds.
In this introduction we cannot attempt to discuss these complicated reaction mechanisms in detail. In the case of thermic reactions our knowledge is particularly limited. Many different reaction pathways and byproducts are known, but there are more hypotheses and corresponding conclusions than with other aroma formation mechanisms. Some special aspects only will be discussed here.
As mentioned, aroma formation by the non-enzymatic browning reaction depends on the presence of enough free amino acids and sugars or peptides. An indirect proof of this fact was recently published with research about the prediction of optimum roasting conditions for peanuts. By means of mathematical models using free amino acids and sugars as variables, C. OUPADISSAKOON and C. T. YOUNG (1984) found a high correlation with several sensory attributes.
An important group of aroma components formed during such high-temperature roasting are pyrazine derivatives. In figure 55 the relationship between the content of cocoa beans in free amino acids, reducing sugars and pyrazines is shown. While the two reactants decrease with increasing roasting time, the pyrazine content rises up to a limit value of approximately 1 mg per 100 g. This prompted the suggestion that methylated pyrazines might be used as an index for optimum roasting degree of cocoa beans (G. ZIEGLEDER 1982). Alkyl and methoxy pyrazine derivatives

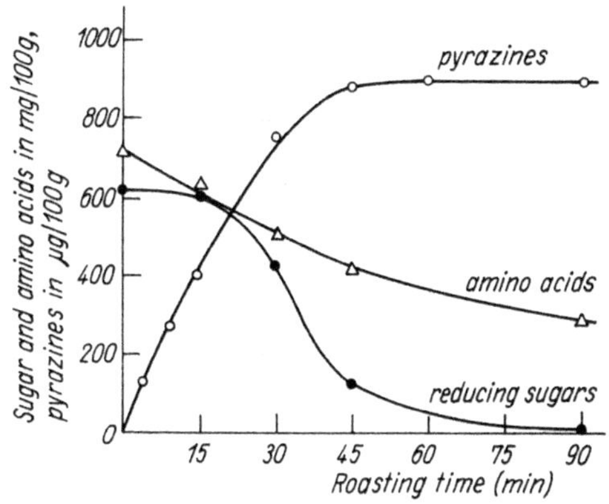

Fig. 55
Relationship between the degradation of free amino acids and sugars and the formation of pyrazines during roasting of cocoa beans
(G. A. REINECCIUS 1972)

are claimed in some patents to substitute cocoa partially, in special formulae.
One of the principal points of our knowledge is the fact that free amino acids and peptides are the limiting factors within most of the food browning systems. As a consequence, proteases must have a positive effect on the aroma production later carried out during heating. Different pathways of amino acid transformation into aroma components are summarized in figure 56.
Aldehydes are one of the main products of amino acid degradation in the course of the MAILLARD reaction. Some of the free amino acids are precursors of a process which is known as STRECKER degradation. According to the scheme shown in figure 57, the amino acid in the first step is reacted with diketone groups formed from the sugar moiety. At the end of the reaction chain the amino acid has been transformed via an oxidative desamination into an aldehyde with one less carbon atom (A. SCHÖNBERG et al. 1948, A. SCHÖNBERG and R. MOUBACHER 1952).

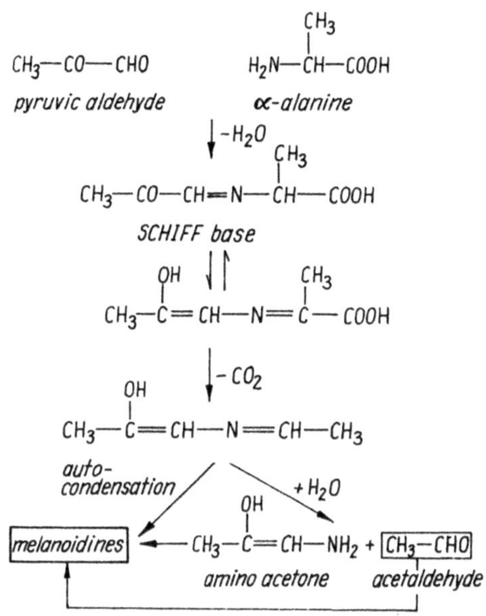

Fig. 57
STRECKER degradation of amino acids during non-enzymatic browning

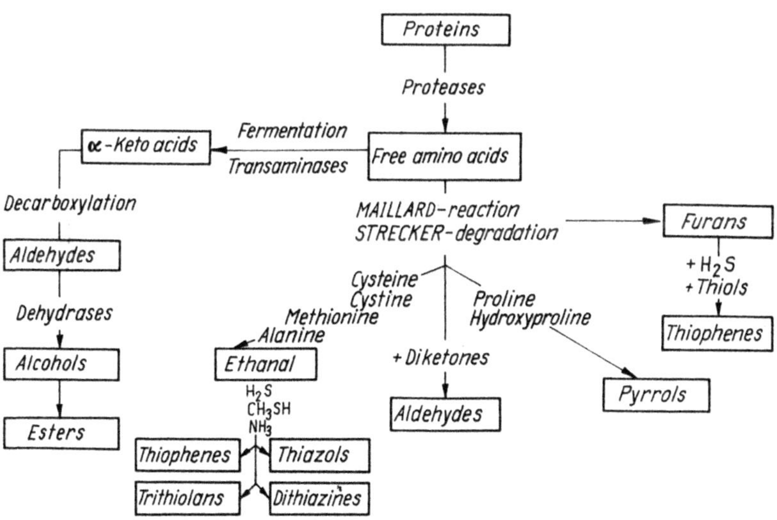

Fig. 56
Different pathways of amino acid transformation into aroma compounds

Table 33
Free amino acids as additives to rye bread and their effect on crust flavour (0,3 g amino acid per 100 g dry matter added)

Added amino acid	Flavour change of the crust	Increase in STRECKER aldehydes (ppm)		
		type	crust without additive	crust with additive
α-alanine	intensive malty note, low bitterness	acetaldehyde	16	28
phenylalanine	malty, floral, honey-like	phenylacetaldehyde	4	17
valine	strongly bitter, nut-like, malty	2-methylpropanal	16	129
leucine	bitter, malty, aldehyde-like	2-methylbutanal	10	61

Table 33 summarizes those amino acids and aldehydes which produce typical and intensive aroma effects in the bread crust after being added to wheaten bread in the dough phase. The test was carried out in such a way that at the end of the dough fermentation phase a small part of the dough was separated. After adding one single amino acid to this small part and mixing, this supplemented dough was stuck over the surface of one part of the bread loaf. After baking, the two parts of the crust were different: the supplemented part showed more intensive browning, a higher amount of STRECKER aldehydes and typical flavour changes. The latter were, however, unbalanced for each amino acid. This shows the complex nature of natural bread aroma, which is based on many different amino acids or STRECKER aldehydes, together with aroma components of other types.

Figure 58 demonstrates the effect of leucine supplementation in rye bread on the aldehyde spectrum demonstrated by thin layer chromatography. A characteristic increase in two components can be observed: furfural as a reaction product of the sugar moiety, and 3-methyl butanal as the STRECKER aldehyde produced from leucine during baking. The flavour change of the bread crust as shown in table 33 must be caused partially by 3-methyl butanal, because this aldehyde is known to have high aroma effectiveness (i. e. a very low threshold value; M. ROTHE et al. 1972).

As a third group of aroma-forming mechanisms, an **oxidative aroma formation** is mentioned in figure 49. Lipid oxidation is the main reaction in this field. It must be of some interest that our earlier opinion of fat oxidation products having only negative influence on the flavour properties of foods has changed somewhat in recent years. Lipid off-flavour is only one part of oxidative aroma. Today we know that special alkenals and alkadienals, as well as methyl ketones in many cases, have a positive key character for some foods (see figure 59). Moreover, lipid peroxides are now known to be precursors for positive aroma components (F. DRAWERT et al. 1966, C. ERIKSSON 1975, P. VAROQUAUX and C. AVISSE 1975, W. GROSCH 1982).

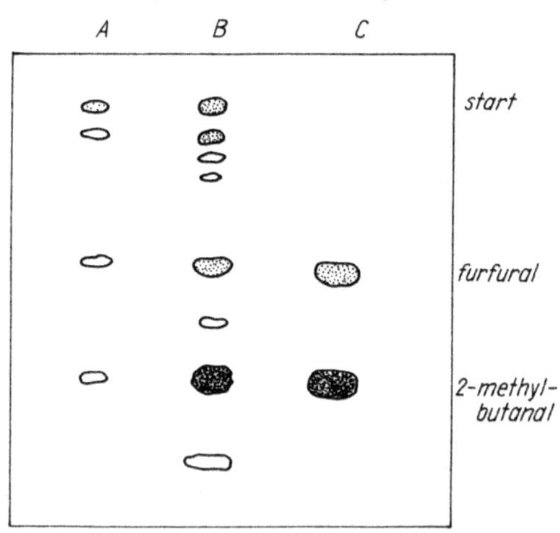

Fig. 58
Thin layer chromatogram of volatile carbonyl compounds isolated from rye bread crust and the effect of added leucine as precursor
(2,4-dinitrophenylhydrazones on Kieselgel G; A = crust without additive; B = crust with 0,1 % leucine added to the dough; C = test substances)

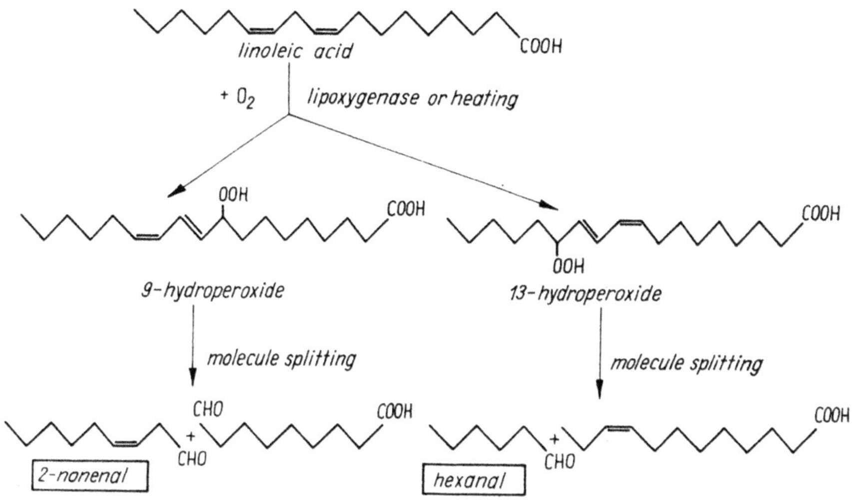

Fig. 59
Formation of volatile aldehydes via peroxidation of polyenoic acids

Peroxides as aroma precursors are included in special types of MAILLARD reaction, too. Thus chicken flavour and fried fish flavour seem to originate from a secondary reaction between peroxides and products of protein hydrolysis (J. POKORNY et al. 1973, 1976a, 1976b, R. SCHRÖDTER et al. 1983, W. GROSCH et al. 1975).

3.2. Interrelation between practical aroma problems

Thus the basic principles of aroma formation can be reduced generally to a few reaction mechanisms. However, the considerable differences between the sensory effects of different foods, as well as between their qualitative and quantitative aroma spectrum, indicate that obviously the same basic reaction can cause highly differentiated aroma sensations. Reaction velocity and pathways of aroma formation seem to depend not only on the nutrient composition of the food source but also on milieu parameters and conditions like pH-value or aroma-binding capacity which change from food to food. Consequently, the chemistry and biochemistry of aroma formation are of a general scientific character only, and cannot be adopted generally for solving aroma problems which arise in practice.

From the practical point of view, another classification principle is also successful. Practical problems of food aroma are mainly concentrated on three directions, i. e. **aroma formation, aroma retention** and **aromatization**. Figure 60 divides the field of practical aroma research into these three central aspects.

Under the term 'aroma formation' we include the large group of foods where the aroma is formed during processing (secondary aroma). 'Aroma retention' is the most important problem with all those foods where the aroma is originally present in the raw material (primary aroma). Aromatization' or 'flavouring' characterizes the traditional field of aroma and flavour research, dealing with flavour-rich spices and aroma concentrates to be added to types of foods and beverages which are naturally poor in, or free of, flavour and aroma. In this case the final products need such additives to acquire normal flavour quality and palatability.

In figure 60 foods with similar aroma problems are arranged next to one another. As mentioned earlier, enzymatic and thermic formation mechanisms are possible within the largest group of 'process flavours'. Sour milk products, cheeses, butter and wine are examples of a microbial or enzymatic aroma formation. On the other hand, meat, fish or coffee

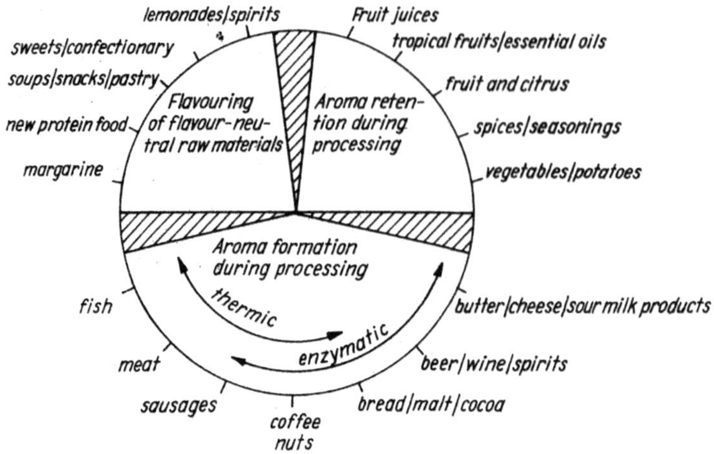

Fig. 60
Cross-links between practical aroma problems of different foods

obtain their typical aroma in the course of thermic processes like cooking, roasting or frying. Frequently both types of reactions overlap or supplement each other. Often, however, enzymatic and thermic processes are concentrated in separate steps of the processing technique. Thus parts of the aroma of bread or of cocoa are produced during the fermentation step. The aroma spectrum is completed in these cases, however, by supplementation with aroma components which are produced in a second thermic step mainly by non-enzymatic browning reactions.

The **aroma retention** of aroma components

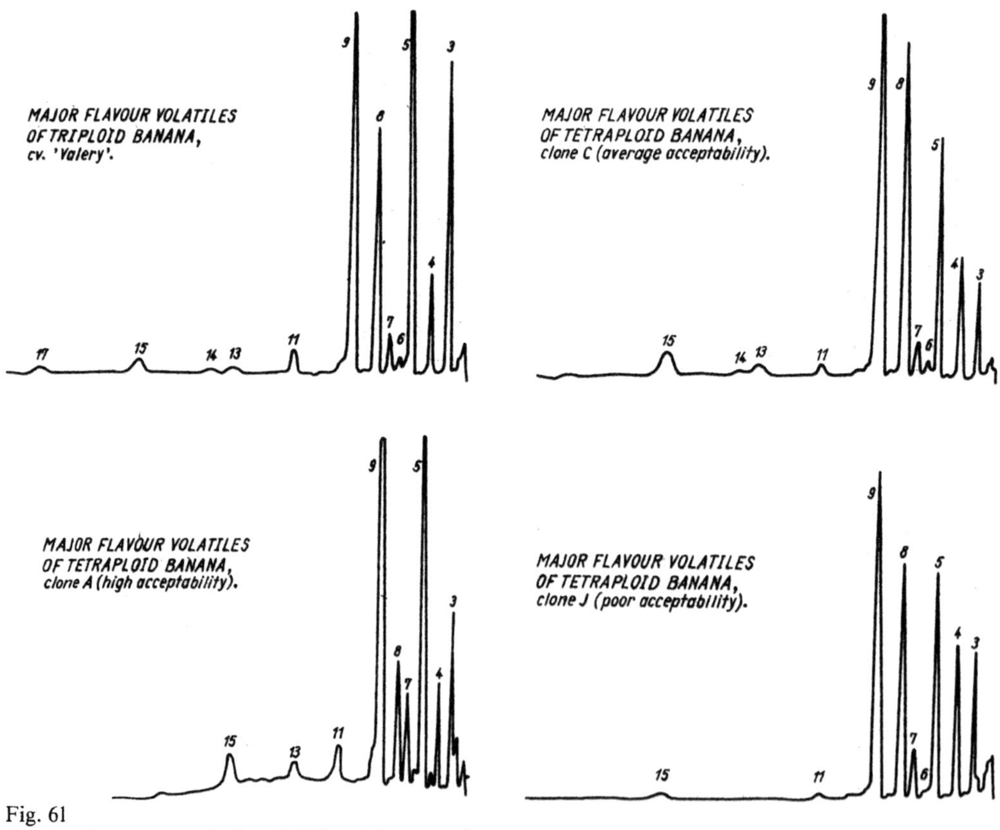

Fig. 61
Dominating aroma volatiles of different banana clones (J. Baldry 1982)

naturally occurring in fruits and vegetables originates from the amount present in the raw material. This value may be influenced by environmental and climatic factors as well as by breeding procedures. As an example of this influence, figure 61 compares the main volatiles of new tetraploid banana clones. Obviously the acceptability depends on the amount of volatiles (J. BALDRY 1982).
Because of the high moisture concentration occurring in most fruits and vegetables, concentration, drying and sterilization processes are of central importance for this group of foods. The removal of water is an economic and transport problem, but also a flavour quality problem. Different techniques can be used, some of which have high costs but good aroma retention (lyophilization) and others which have low costs but involve aroma losses and changes (see also figures 17 and 62). Aroma binding problems are included here, for example the selection and use of appropriate carrier substances during spray drying. An example for aroma problems connected with the technical equipment are an optimum fractionation of aroma concentrates in the process of juice concentration (F. EMCH 1967) or of smoke concentrates which are free of carcinogenic components.

Aromatization or **flavouring** of food products which are poor in, or free of, a flavour of their own, is the third large group of practical aroma problems. All flavourings for alcoholic and non-alcoholic beverages, sweets and confectionery, snacks, soups, puddings or margarine belong to this field, which deals with the production as well as the application of such additives. Natural and nature-identical aroma concentrates are of most interest, while synthetically produced 'artificial' flavourings are increasingly being replaced by others. The increasing trend towards nature-identical products in particular is caused by the steadily increasing prices of most natural flavourings and flavour-rich foods, which have mainly to be imported from tropical countries.
An optimum composition of flavourings from hundreds of single aroma components, essential oils or other aroma complexes, remains the most important problem for the flavour industry. The steady improvement of newly developed or composed flavourings and aroma concentrates gradually brings their quality closer to that of the more expensive natural products: there is therefore more change of success for new flavourings. However, legal and toxicological problems are relevant

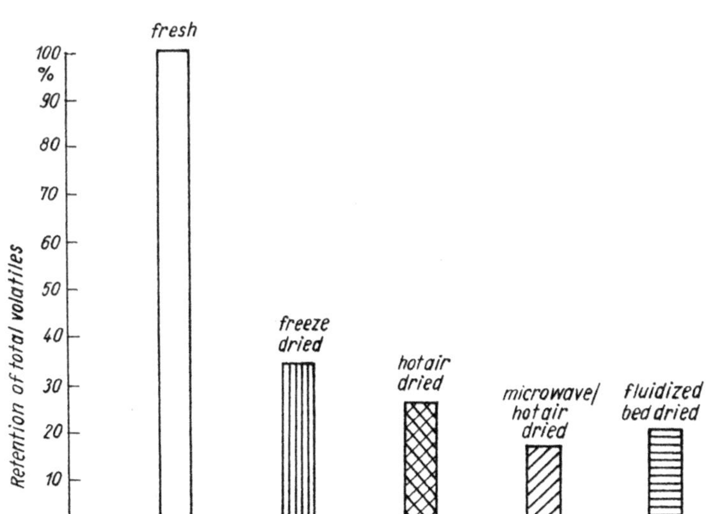

Fig. 62
Losses in carrot volatiles caused by different drying techniques (E. KAMINSKI et al. 1985)

here and must be considered. Another important point in this whole area is the need to carry out more consumer tests, in order to increase our knowledge of acceptance factors.

Comparing aroma problems within the three groups mentioned, only few similarities and parallels can be found between foods arranged side by side in figure 60. As well as the previously mentioned principal problems of food flavouring, cross-links exist particularly in the 'aroma formation' group. Here aroma-forming reactions are often based on the same principle: for example, the formation of aroma components by fermentation and microbial metabolism, respectively, or the widespread non-enzymatic browning reactions. Differences in the composition of the medium change the system in such a way, however, that parallels in aroma problems of different foods still remain exceptional. The specifities of the various aroma spectra are high enough to prevent the discovery of general guidelines for aroma production or aroma control. Though research in the field of cross-links between different food flavours continues, most practical aroma problems are highly specific. Thus most of the technical production parameters related to formation or retention or aroma control methods have to be adapted to the special food item under consideration.

3.3. Nutritional aspects of food flavour

The progress in food science and technology achieved within the last century, as well as extensive social changes, have influenced human nutrition much more than ever before in history. Typical alterations in diet are still continuing: they are more advanced in industrialized areas than in developing countries. A large number of the young nations of Africa and Asia, founded in this century, are nowadays in the phase most of the highly developed countries passed through decades ago.

A characteristic effect accompanying the general improvement of human living conditions is the increase in variety of food and delicacies available, followed by overnutrition and malnutrition of large parts of the population. This is directly related to the desire of man for a higher and improved quality of life in all areas, including the flavour quality of foods. The wide choice of different foods available, which is evident today in highly developed countries, has a strong influence on the amount of food consumed and can thus be the cause of overnutrition and overweight.

This is, however, not a problem only of the twentieth century. Because of typical social developments, luxurious nutrition is nearly as old as human cultural history. This type of nutrition was, however, limited for many centuries to the privileged classes. The courtiers of ancient Egypt, as well as the rich classes in Greece or Rome, enjoyed a luxurious existence with special refined foods and meals: we can tell this from descriptions of foods and dishes as well as from pictures showing overnourished people. At the same time, the main part of the population was supplied only with food which was easy to produce and cheap. Soldiers on the march in these societies are an example: normally cereal grains were their only foodstuff.

These products usually have a low flavour intensity, and therefore underprivileged people had an interest in changing this situation. There were two choices: on one hand, attempts were made to improve the low-flavoured staple foods available for these classes of people. On the other hand, there was a sustained interest in achieving or imitating the consumption habits of the higher classes. Several of the historical conflicts and fights between different classes can be explained by social differences and discrimination of this kind (fig. 63).

Considering the present-day food deficiency conditions in the world, hitherto only overcome by highly developed countries, we can find trends which seem to be related to the problems mentioned above. There are two characteristic phases which are connected with flavour aspects. In the first phase the nutritional deficiencies of large parts of the population must be overcome This target, already attained in most of the industrialized coun-

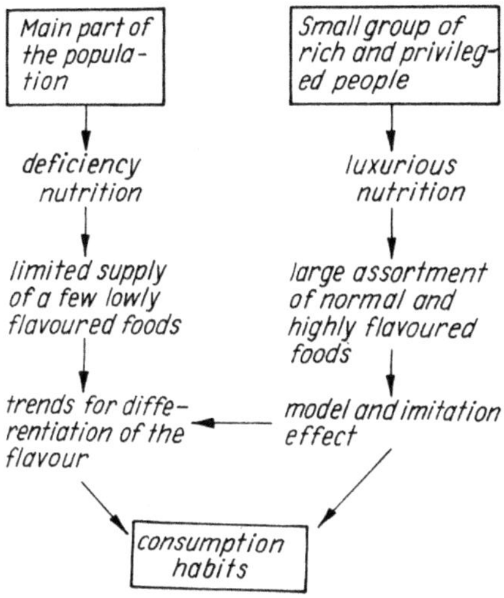

Fig. 63
Social aspects as a source of establishing consumer habits

tries, is generally reached by an increase in agricultural production. Consequently, in all developing countries today staple foods of plant origin take first place.

The flavour intensity of these products is small. A sufficient caloric food supply is, however, more important in this phase of country's development. Naturally the overcoming of hunger is a much more critical social and dietary problem for the consumer than the optimum flavour quality of the foods available.

Figure 64 demonstrates such effects by calculating tabulated FAO data (FAO Production Yearbook 1972). Each column represents the average caloric value for 10 countries in one year chosen from the period between 1950 and 1970. The years and countries included in the calculation had an average caloric supply corresponding to the caloric values mentioned in the figure. The columns demonstrate that parallel to the enhancement of the caloric supply a decrease in the percentage of low-flavoured cereals and/or potatoes can be observed within the total palette of foods. This trend is particularly noticeable if the average caloric consumption corresponds to the normal daily value of approximately 3000 kcal (= 12550 J) per head per day.

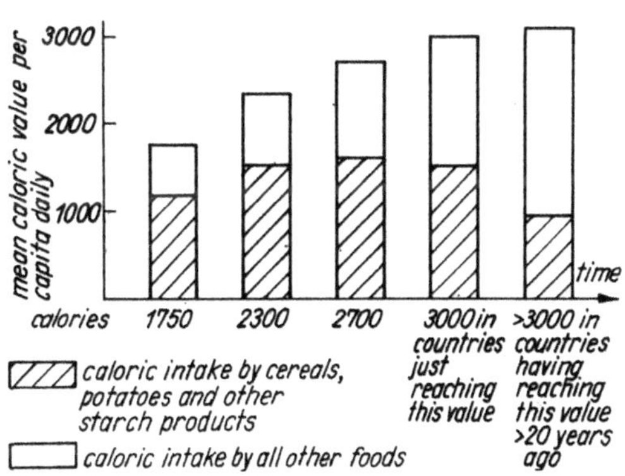

Fig. 64
Percentage of low-flavoured cereal and potato products in total caloric intake in different phases of countries' development (M. Rothe 1977)

Figure 65 represents an example of undernutrition and hunger on the right, and of overnutrition of average people on the left. These two extreme situations are characterized not only by too low or too high a supply of nutrients. A parallel change in composition of the different nutrients can be registered. As shown in figure 65, overnutrition indicates simultaneously excessive amounts of fat rather than carbohydrates in the diet. In contrast, developing countries with undercaloric nutrition show an excessive supply of carbohydrates. This is caused by unbalanced nutrition with large amounts of cereal products in the diet. Here protein and fat remain far below the recommended levels. Both types of nutrition can be discussed as special forms of malnutrition.

The typical trend in the consumption habits of highly developed countries can easily be demonstrated by the situation in the two German states in recent decades. Here, as a consequence of food deficiency in the years following World War II, nutrition was reduced to a level characteristic of developing countries. This situation, however, changed within a relatively short time to typical overnutrition with high fat and sugar consumption. As in other highly developed countries, intake is now far above the recommended levels (Table 34).

From the nutritional point of view, this trend appears to be very unfavourable. High amounts of fat and sugar in the diet accelerate overweight, which is a widespread pheno-

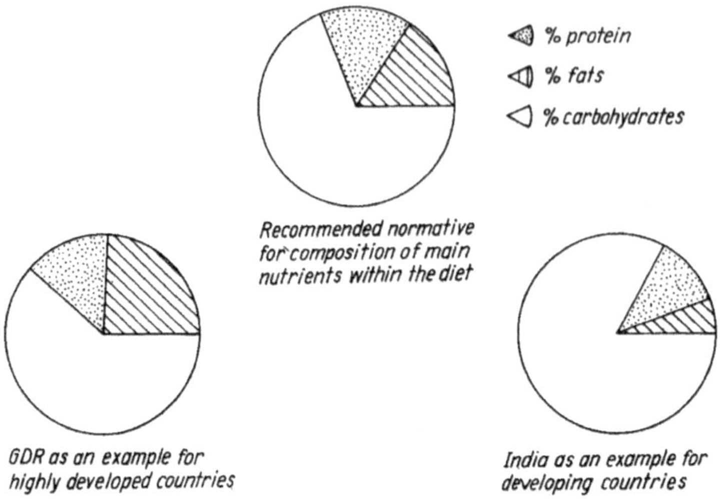

Fig. 65
Most important forms of malnutrition compared with the recommended diet

menon in all industrialized countries. For most European countries, as well as the USA and Canada, between 25 and 40% of the adult population are considered to be obese (more than 20% overweight), females rather more than males (K. VETTER 1973). In children of school age, the percentage of those overweight has also started to increase.

Overweight promotes several diseases, of the coronary system or the intestine, for example. Figure 66 shows the risk for several diseases for overweight people compared with persons of normal weight. If the current trend continues, in less than a decade from now a third of the adult population will require a special diet in order to reduce these health risks.

The flavour and aroma or our foods are undoubtedly one of the reasons for this. It is not the fat or the sugar itself which induces the consumer to eat more and more food products which are rich in these two nutrients. Obviously a more pleasant flavour and aroma seems to be responsible for high consumption rates. If we compare the quality of

- milk and skim milk,
- ice cream with different fat content,
- full-fat cheese with low-fat cheese,
- normal and fat-reduced sausage,
- normal confectionery with that of reduced fat and sugar content,
- canned fruits with and without added sugar

in every case we will feel that the fat-rich or sugar-rich product is more attractive and more widely acceptable.

If we exclude special strict dietetic regimes, there is no food or meal which we consume

Table 34
Trends in nutrient supply in the two German states (g per caput and day)

Recommended	Protein		Lipid		Carbohydrate		Sugar	
	74		82		314		—	
Consumed	GDR	FRG	GDR	FRG	GDR	FRG	GDR	FRG
Year								
1955	71	79	102	120	410	408	75	78
1960	75	79	119	126	395	383	80	83
1965	79	80	127	131	384	356	83	88
1970	82	82	130	140	380	350	94	93
1973	82	83	133	140	383	349	100	92
1982	82	86	126	143	342	350	92	92

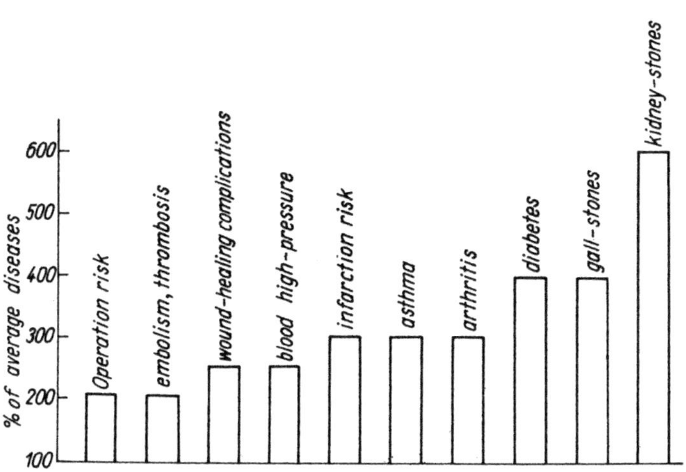

Fig. 66
Disease risk of overweight persons compared with the risk of normal-weight (K. VETTER 1973)

because of its nutrients only. Moreover, food with poor aroma is often refused (J. SOLMS 1973). Foods based on a neutral raw material like sweets, puddings, lemonades or margarine are not attractive to the consumer before aromatization. Foods and meals with low flavour intensity such as milk, bread or potatoes are normally consumed along with, or in combination with, flavour-rich supplements. As an example, the rice meals usually consumed in south and south-east Asia need high amounts of spices and/or glutamate (soya sauce) as a supplement, especially when not enough meat is available.

It depends on traditional and cultural customs whether a flavour or aroma is registered as more or less pleasant, or whether a food is refused because of an extreme flavour quality or intensity. Absolutely new and unknown flavours are normally not easy to introduce, because consumption habits are quite strongly anchored. H. GLATZEL (1968) observed that even in hunger situations people refused food with an unknown flavour.

The influence of food aroma quality on the consumption rate can be demonstrated by the amount of dietetic products consumed by patients suffering from special diseases. Such foods are characterized in most cases by decreased flavour and aroma quality. As a consequence, most patients use the normal (fat-rich, sugar-rich or salt-containing) food products. In the 1970s, for example, sorbitol was the sweetener mainly used in the diet of diabetics in the GDR. Many special dietetic products were brought into the market for these people, but their poor flavour caused very low consumption rates. The consumption of sugar-reduced bakery and confectionery products, for example, did not exceed 10% of the average amount that normally healthy people consumed in this food group.

From such examples it becomes evident that flavour and aroma greatly stimulate food consumption. As a consequence, a typical trend towards foods characterized by high flavour quality is observed in highly developed countries. For the consumer, an increasing standard of life is equated with a desire for more differentiation in the flavour of the diet and thus with increased consumption of flavour-rich foods. Statistically registered consumption rates for food types high and low in flavour prove this thesis as shown for example for the GDR in table 35 and figure 67.

Figure 67 indicates that in the case of four selected groups of flavour-intensive foods and beverages a relationship with high correlation coefficients can be observed. The decrease in the consumption rates for rye bread, with relatively poor flavour, was accompanied by greatly increased rates for meat, sugar, cocoa or beer which are all characterized by a much higher flavour intensity. Each point on the

Table 35
Consumption trends in food with high and low aroma intensity in the GDR
(kg per caput and year; Statistisches Jahrbuch der DDR 1986)

	1955	1960	1965	1970	1975	1985
Coffee	0.3	1.1	1.8	2.2	2.4	3.5
Tea (g)	77	88	90	91	101	139
Cocoa products	1.1	1.9	2.1	2.4	3.0	42
Sugar and sweets[1]	27.4	29.3	30.1	34.4	36.8	39.6
Chicken	3.4	3.7	4.0	5.1	7.6	10.2
Beer (l)	68.5	79.5	80.6	95.7	119.7	141.6
Spirits (l)	4.4	3.5	4.7	6.6	8.6	15.2
Flour and pastry	121.6	101.6	100.2	97.3	94.8	99.3
Rye bread[2]	68.3	49.9	43.8	40.4	35.3	35.6
Potatoes	174.6	173.9	156.5	153.5	142.1	143.4

[1] sugar base
[2] flour base

figure demonstrates the situation for one year in the period between 1955 and 1980. The curvillinear shape of the relationship, however, indicates that the obvious exchange within the diet will not continue in the future and may approximate to a minimum limit in rye bread consumption. This minimum amount of 30 kg rye bread per head and year may be maintained in the diet for sensory or physiological reasons.

Another related effect should be mentioned here. Obviously the consumption in low-flavour foods can be stimulated by accelerating the flavour intensity significantly. Thus the small increase in the consumption rate of potatoes in the most recent period mentioned in table 35 seems to depend on the appearance of potato specialities on the GDR market during this time (potato chips, French fries in prepared form, instantized potato puree). A similar effect may contribute to the turning point of the bread consumption curve which can be registered in some European countries: the appearance of speciality breads on

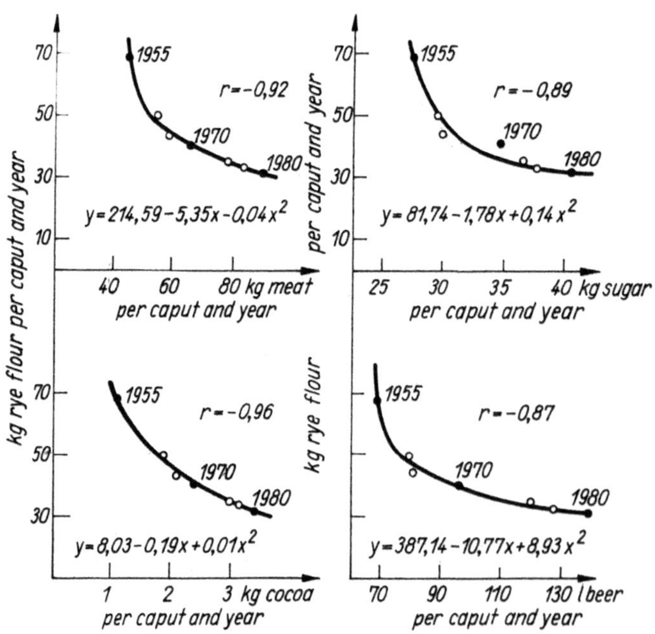

Fig. 67
Negative correlation between the consumption of rye bread and of some flavour-intensive foods in GDR

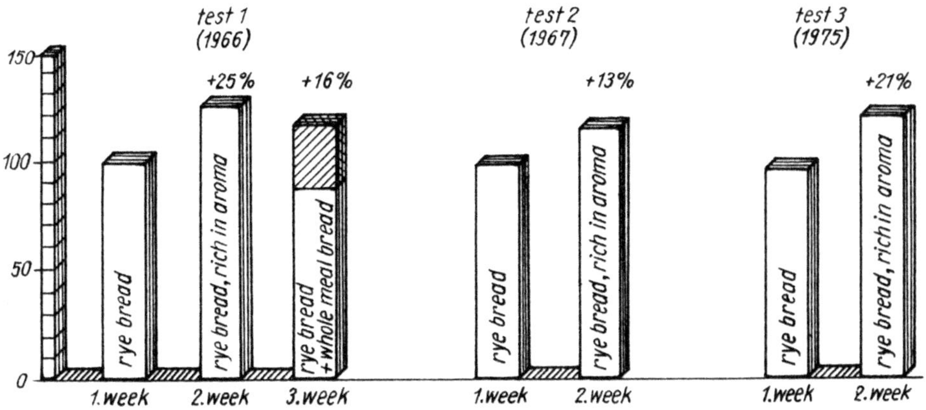

Fig. 68
Bread consumption rates influenced by bread aroma intensity
(M. ROTHE and G. REHFELD 1967)

the market stimulates the consumption by means of their more intensive and attractive flavour.

Effects of such type have been proved two decades ago by consumer tests. Some results are summarized in figure 68; they indicate that the enhancement of the aroma quality or intensity increases bread consumption (M. ROTHE and G. REHFELD 1967). The test was carried out in such a way that in boarding establishments for adults the amounts of bread consumed per head per day were registered. All other foods and meals were offered in unlimited amounts. From week to week, bread type and bread quality were changed. Rye bread was used, being the kind mainly consumed in the GDR.

According to figure 68, the supply of aroma-rich bread caused an increase in bread consumption dependent on the aroma quality. The same was true when aroma-intensive types such as whole meal bread were offered in addition. Though the dark bread was only partially consumed, the bread consumption increased in total. Supposing that the whole energic (caloric) intake does not change, a higher consumption rate of bread instead of the fat-rich spreads can be expected. This fact is of interest not only from the nutritional point of view but also for the bread industry: as a consequence, more varieties of bread and more bread specialities with improved and intensified flavour have appeared on the European market in the recent years (W. SEIBEL and H. STEPHAN 1970).

Another type of consumer test was arranged by our research group with similar methods and aim. For a period of several weeks groups of test people were given a breakfast of bread and rolls under controlled conditions. They were instructed to use butter and different types of spreads in their individually accustomed way, and to weigh all slices without and with spreads. Cheeses, sausages and jam with varying flavour intensity were used. For every two-week period the combinations remained unchanged.

Selected results are shown in figure 69. Depending on the flavour intensity, the amount of spread related to a constant bread weight varied. Simultaneously, the ratio of fat or sugar content of the spread to bread weight was changed. In the case of Camembert cheese a highly significant difference was observed between the unripe product and the same cheese in the fully ripened and aroma-rich state. In all other cases only trends were registered. The aroma intensity of the spreads, however, seems to change the composition of the meal consumed in a positive way, when considered from the nutritional point of view.

Consequently, an improved or intensified flavour quality can significantly influence the nutrient spectrum of the daily food intake. This indicates an opportunity to reduce the

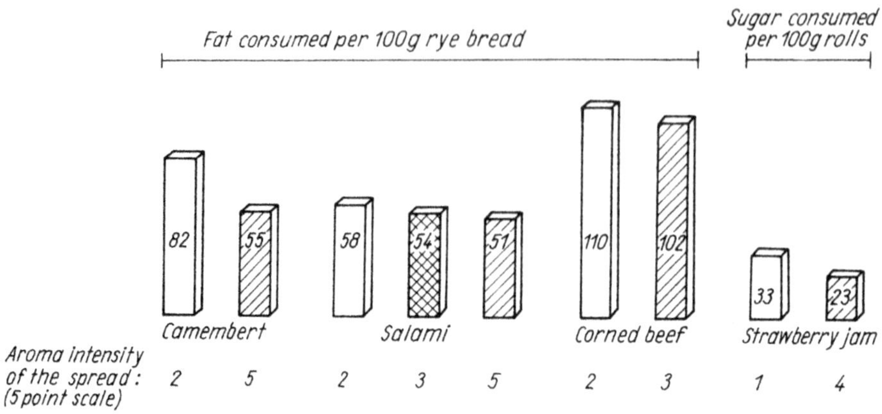

Fig. 69
Influence of the aroma intensity of bread spreads rich in fat or sucrose on the amount consumed with bread (average consumption of included fat or sugar per 100 g bread)

amounts of fat and sugar causing malnutrition in highly developed countries and populations, with their negative health effects. On the other hand, such calculations have to consider the fact that increased aroma intensity stimulates the appetite and may itself be one of the reasons for overnutrition and overweight.

L. HALL and E. J. MERWIN (1981) discuss an evolutionary role of flavours instead of the intentional role characteristic of the early stages of human development. In their opinion, flavour was a most important aid for the primitive hunter or gatherer, encouraging the supply of nutrients essential for survival. Thus a series of taste preferences evolved, probably genetically based, which we still retain. They may partly determine our preference for sweet products, which can be observed even in new-born babies. Pure solutions of sucrose are consumed by new-born children in higher amounts than unsweetened ones, even when the latter have an optimum nutrient composition (J. DESOR 1976). Children generally prefer higher sweetness in food than adults do. The same is true of saltiness. Maternal preferences for intensive sweetness is not transmitted to the new-born, however; this has been shown by recent experiments with rats (C. L. THEALL and W. L. CHENOWETH 1983).

H. W. MOSKOWITZ et al. (1975) assume that dietary experience also has an influence. They found in a comparison between Indian labourers and Indian medical students that the latter evaluated relatively low concentrations of bitter or sour substances as unpleasant. The workers, in contrast, rated increasing amounts of the same components as more and more pleasant. Experience factors must also be the reason for an effect observed by G. K. BEAUCHAMPS and M. MORAN (1982). In studies on new-born infants, children who were fed sugar solutions very early by the mother were found to consume higher amounts of sugar at the age of six months or two years. When the early sugar feeding period was very short, however, the behaviour of the child remained unchanged.

When a long-term reduction in special taste substances becomes necessary because of dietary regimes, the intensity ratings can change significantly. This has been found to be valid for the reduction in sodium chloride and the salty taste (M. BERTINO et al. 1982), as well as for diabetics who are not allowed to consume sugar. The latter group consequently register sweet products as more intensive in the sweet taste than normal consumers do.

Studies of different cultural groups finally lead to the conclusion that there is some variability originating in different cultural habits. Thus Taiwanese-born Americans were found to rate cookies as sweeter than Americans of European ancestry: they also show a different appreciation of salty products (M. BERTINO et al. 1983). Differences were also

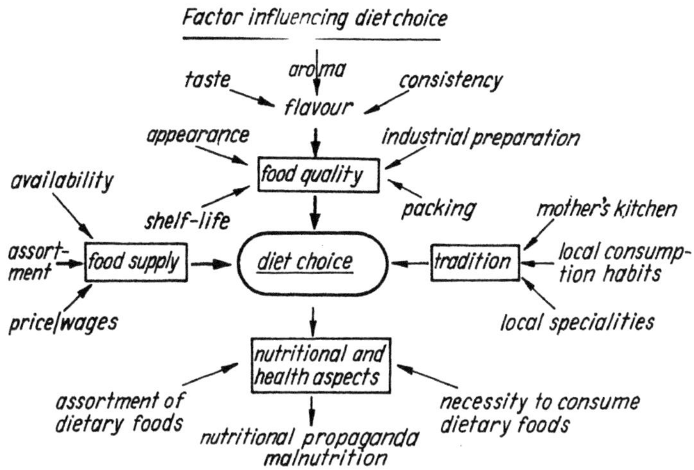

Fig. 70
Factors influencing diet choice

registered between black and caucasian children, with blacks showing preferences for higher concentrations of sucrose, sodium chloride and lactose (J. A. Desor et al. 1975).

Another interesting but so far unproven point of discussion concerns the higher taste intensity produced by most bitter-tasting substances compared with sweet, sour or salty compounds. Considering the bitter taste of numerous natural occurring toxic components, the supposition is obvious that tne higher sensitivity to bitter substances may be genetically fixed as a danger signal.

The acceptance of flavour in foods is influenced by the nutrient directly. A fibre-rich diet of long duration reduces sensitivity to aroma and taste substances (K. H. Plattig and G. Berg 1974). D. H. Cohenour and D. H. Calloway (1969) tested a large spectrum of aroma concentrates for their acceptance in special dietetic regimes. Groups of test people were selected who were provided with a long-term diet rich in carbohydrates, protein or fats. They had to rank the different aroma concentrates as to their palatability. As a result, protein-rich diets were accepted for a long time in sweetened form only. Of the numerous aroma concentrates tested, only those with citrus character were accepted in all types of diet.

Overweight children and adults show the same sensitivity to flavour perception, but they have an aversion to strongly sweetened beverages (J. A. Grinker 1975).

The examples discussed show that many factors influence food choice, and that they depend on food flavour. Figure 70 summarizes some aspects. Besides the food quality, which is a central point, there are three groups of influence factors: tradition, food supply, and nutritional or health aspects. All these factors are surpassed and dominated by the food flavour quality. Modern consumers are guided in their food choice by flavour and aroma. When we accept this point we should also consider that there is an opportunity to overcome the malnutrition and overnutrition currently widespread in highly developed countries in a new way: by improving the flavour of such foods whose consumption has to be increased from the health and nutritional point of view! That means, for instance, better aroma in dietetic products, as well as in foods which are low in fat and sugar content.

4. Future aspects

The rapid development of aroma research in the last 30 years is a consequence of new equipment and instrumental techniques for the analysis of trace volatiles, as well as of great progress in food technology and food production. If we accept the argument that the sensory properties of our foods and meals stimulate the spectrum of food consumed, we must pay more attention to this factor. The importance of aroma problems is therefore bound to increase in the foreseeable future.

There are various new tasks and new perspectives for many branches of the food industry. In recent decades we have become aware that many traditional and sometimes old-fashioned processes in the food industry have been superseded by new, more rationalized and more economic food production. However, if this trend continues, with increasing remoteness from the traditional processes, losses in aroma intensity and/or aroma quality will often result. Here we find an important field of potential interest for consumer acceptance and nutritional aspects.

Agriculture and breeding research must also be included to this discussion. Fruits and vegetables owe their increasing interest to the consumer mainly to their flavour properties, but the breeding aims are never focused on flavour and aroma. The main aims of breeding research are normally economic factors like the yield under different climatic and environmental conditions, the resistence to infection, or good processing and marketing properties. Flavour and aroma should receive much more attention in future. We have to remember such warning examples as that of cultivated giant-size tomatoes which were lacking in aroma; they were soon refused by consumers, and have caused a decrease in consumption (H. HULPKE 1969).

Tomatoes are also an example of the strong influence of the fruit-ripening process on flavour quality, which is not sufficiently considered in modern large-scale fruit production. A. A. KADER et al. (1977) reported on tomatoes picked at earlier stages of ripening, as well as those artificially ripened at 20 °C. They observed changes in the normal sour/sweet ratio, as well as less tomato-like aroma and more off-flavour. Recently this was again demonstrated by F. HAYASE et al. (1984), who found large differences in the quantitative aroma spectrum varying with the ripening conditions.

Some publications in this field also indicate a change in thinking about the most important food quality factors. As an example figure 71 demonstrates head space gas chromatograms of two varieties of passion fruit, with obvious differences in the content of volatiles. In practice, however, *Passiflora edulis*, which has a much higher content of aroma compounds, was replaced in Australia by *Passiflora flavicarpa* and some hybrids which have higher crop yields and are more resistent to viral attack. Now the authors ask, with the voice of the consumer, why passion fruit no longer tastes as it used to (F. B. WHITFIELD 1982).

The continuous replacement of older food technologies will be accompanied in the future by production and use of new food sources and materials. This development is hastened by the food production problems connected with the rapid increase in world population. The key for solving future nutritional problems in the world is new food resources. We have, however, enough examples of the fact that off-flavour problems inhibit the introduction, for example, of new protein sources of plant origin; the same may be true for microbial proteins. In highly developed countries such substitutes have little chance of entering the market, and therefore there are proposals to use them in developing countries first. But in every case we must find a way for solving these off-flavour problems in future. The term 'tailor-made' for new foods from new sources is today valid only for their main constituents and not for the flavour, which is one of the most important attributes of such products. Thus the removal of the off-flavour components, or their masking by selected aroma concentrates, may be one of the future possibilities.

The outlook for the flavour industry is favourable. The traditional methods of aroma ex-

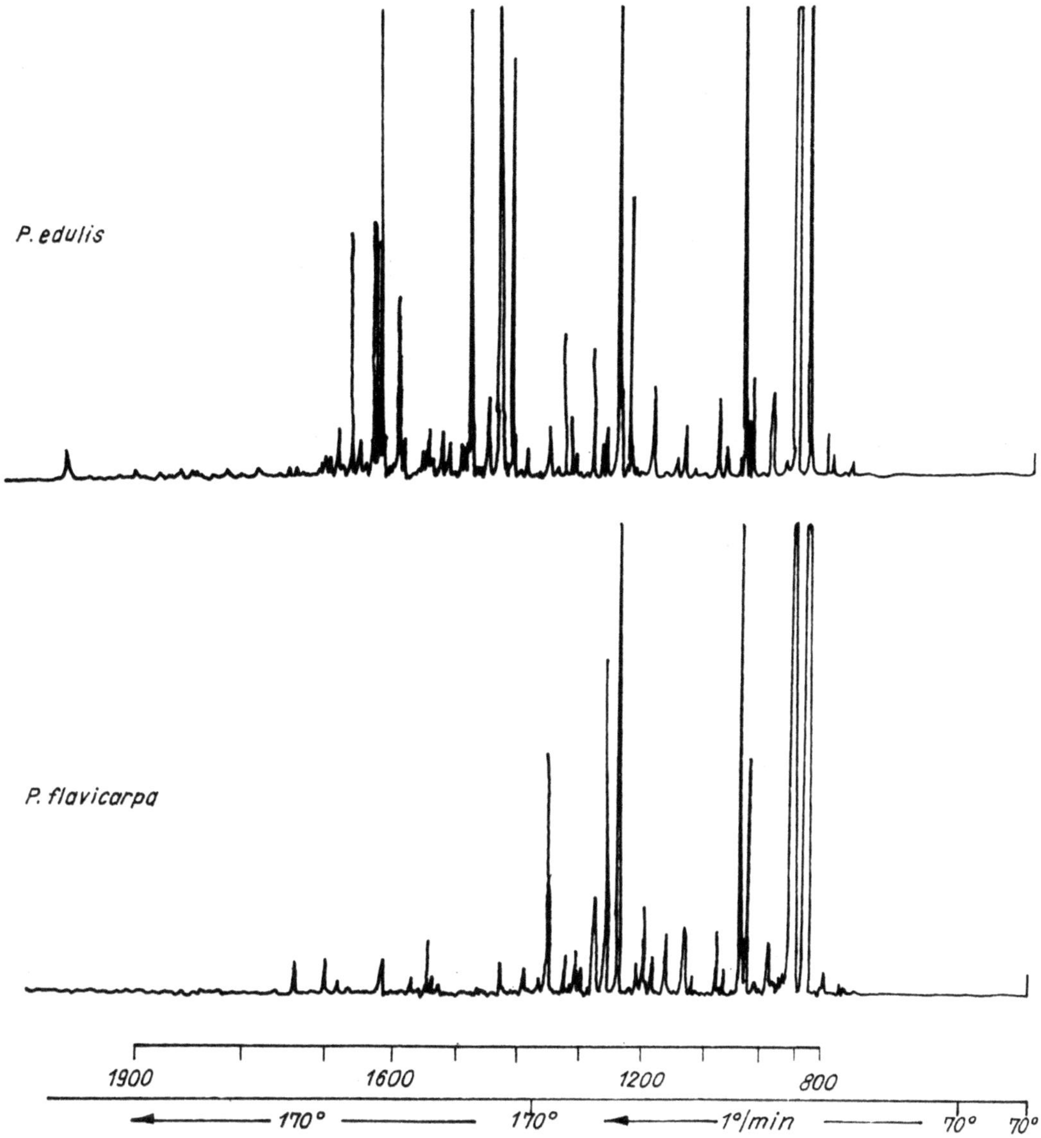

Fig. 71
Head space gas chromatogram from purple highly acceptable passion fruit juice (*Passiflora edulis*) and juice of a yellow less acceptable mutant (*Passiflora flavi-carpa*) (F. B. Whitfield 1982)

traction and distillation from natural sources will be supplemented by new techniques of thermic or enzymatic formation of aroma complexes. Additionally, new principles of biotechnological aroma production, such as bioconversion or microbial synthesis of special aroma components or aroma complexes as well as aroma production via plant tissue cultures now being in the laboratory and pilot scale will have their chance.

The composition of aroma concentrates from single components, which is currently carried out empirically, will in future be directed by statistical programs. As a corollary, aspects of food law and toxicology will have to be considered more carefully, and at an earlier stage, than in the past.

All our efforts to improve and intensify the flavour quality of foods, and to produce new and better aroma and flavour concentrates, can be successful only if it is possible to evaluate the flavour in sufficient detail. This aim will be reached by the development and use of improved sensory and instrumental techniques, and by their application in solving the large problem complexes of aroma formation, aroma retention, aroma characterization and flavouring.

In summary, we have to expect the following topics and problems to be in the centre of aroma research:

- Completing our knowledge of the qualitative composition of food aromas,
- further improvement of instrumental techniques, in combination with computer analysis,
- more research designed to obtain quantitative data in instrumental aroma evaluation,
- the development of further new sensory methods for detailed flavour characterization.
- the development and use of HPLC methods for determining non-volatile flavourings and aroma precursors,
- the weighting of aroma components as to their importance in total flavour perception,
- the detailed elucidation of formation mechanisms for aroma components,
- research on aroma indices to be used in aroma production and retention problems, and in quality control,
- increased application of statistical methods especially for finding correlations between sensory and instrumental data,
- the development of new complex aroma concentrates via biotechnological processes,
- more consumer acceptance tests to increase our knowledge of the acceptability of new foods with new and changed flavour qualities.

The ultimate aim of aroma research should be discussed as a humanistic one:
to contribute to food improvement and to food variability as well as to the solving of world-wide nutritional problems but, first and foremost, to improve the quality of human life.

5. Time-table about the history of production, processing and consumption of aroma-rich foods and flavourings

3500 BC Viticulture and wine processing known in Egypt and Assyria

3000 BC tone pots with reflux hood and sewer pipe have been found in Egypt which may have been used for distillation of fermented material

honey produced from dates known in Mesopotamia. It is used for sweetening purposes as well as for the production of alcoholic beverages and vinegar

2800 BC the Chinese emperor SCHEN-NUNG mentions cinnamon bark as a spice in his herbal
the Sumerians in Mesopotamia are known as breeders of animals including cattle, pigs, sheep and goat. They produce butter and cheese from the milk

2500 BC industrial beer production begins in Babylon. There are breweries with a stock of some thousand ton barrels

2400 BC white cakes made from flour, milk, cheese, oil and spices are consumed in Egypt on special occasions

2300 BC an Egyptian tablet demonstrates the application of cassia and thyme in brewing

1900 BC a Chinese author describes the production of rice wine, including malting and fermentation

1700 BC the kinds of grapes cultured today are already known in Greece; proof of this comes from the grape kernels found in the sediments of Orchomenos

1500 BC 877 formulae are summarized on an Egyptian papyrus roll mentioning anise, cassia, calamus, coriander, caraway, mint, mustard and sesame

1500—1300 BC bee honey is known in Syria, Palestine, Egypt and the kingdom of the Hittites (eastern part of modern Turkey). These fertile areas are described in the Old Testament 21 times as 'the country where milk and honey flow'

1000 BC black pepper is used in India as a single spice as well as combined with curry

950 BC the Bible describes the visit of the Queen of Sheba (South Arabia) to Jerusalem where she brought as a present for King Solomon loads of spices which were unknown there. The wealth of 'Felix Arabia' lasting for centuries was founded on its wide trade in herbs and spices

9th century BC the Iliad and the Odyssey praise the high degree of viticulture in Thrace (Greece/Bulgaria/European Turkey)

776 BC white bread, olives and fish are presented to the guests at the first Olympic Games in Olympia

700 BC a cuneiform library of the Assyrian emperor ASURBANIPAL contains material on the herbs dill, fennel, cardamom, caraway, saffron, sesame and thyme

600 BC Greek trading vessels bring wine stems to Marseilles, where viniculture starts to expand durings subsequent centuries to the central and nothern part of France and later to central Europe. In the loth century monks transmit citiculture to Saxony. The greatest range is reached in Europe in the 15th century: grapes are produced from Spain to Poland. After the Thirty Years War grapes are cultured only in such areas where the climate is favourable for an optimum flavour quality

475 BC HERODOTUS writes about an Egyptian custom of using raw meat and raw fish in salted dry form

340 BC ARISTOTLE claims that taste sensation depends on the presence of liquid. His erroneous statement that the sweetness of substances would be caused by the simultaneous reception of 'warm' and 'moist' sensations is still being discussed in the late Middle Ages

327 BC the commanders of ALEXANDER the GREAT report a special type of Indian reed which 'produces honey without help of the bees'. The army brings loads of pepper and calamus from India to Greece

300 BC the storage of natural ice and snow in hollows boarded with wood is widespread in Greece. HIPPOCRATES recommends its use in combination with honey, fruits and spices because such ice-cream would improve well-being and health
THEOPHRASTUS tries to classify odours and discusses the difference between odour and taste as well as the sorption of odorous components. He deals with the composition of odours, and discovers that odour sensation depends on the medium

221 BC the consumption of black tea in China is significant because of the imposition of a tea tax. An expansion into other areas does not begin before the 6th century AD

171 BC the first commercial bakeries and cake shops are built in Rome under the influence of Greece

100 BC lists of apple varieties are known in southern Europe

50 BC DIOSCURIDES summarizes all spices and herbs known in Greece and Rome in his *materia medica*. Turmeric root appears for the first time; the use of ginger is described in detail

176 AD the Romans collect taxes on all exotic goods entering the port of Alexandria

300 solid sugar is mentioned in India for the first time

637 THEOPHANES reports on the capture of large loads of sugar after the victory over the Persians, as well as precious metals and spices. This explicit mention shows that sugar was a rarity at that time in Byzantium

768 the first hop-gardens are documented in Europe. In contrast to ancient practice, hops are now used for preparing the wort

780 the Chinese poet LU-YÜ describes the tea plant, its culture and processing, the various types of tea preparation and the equipment for this preparation in a three-volume book

9th century orange trees originating from India and East Africa have spread over Arabia, North Africa and Spain. The peel is used as a seasoning. Other citrus fruits arrive in Europe 700 years later

815 the Japanese emperor orders the cultivation of black tea in five of his provinces after Buddhist monks brought him tea seeds from China and were successful in cultivation in Japan

9th and 10th century the preparation of special kinds of cheese starts in Italian and French monasteries: these are later designated according to their origin (Gorgonzola, Roquefort, Parma)

973 during the visit of the Arabian physician and tradesman IBN JAQUB to Germany, cloves are mentioned for the first time

996 Venetian trading vessels transport large loads of sugar from Syria and Egypt to Italy

1000 the Arabian physician IBN SINA (AVICENNA) reports on the consumption of a coffee beverage in Abyssinia

1079 the regular addition of hops in the course of beer brewing is documented

1179 the 'Guilt of Pepperers' is founded in London, the first union of spice tradesmen

12th century during the crusades to Palestine, aromatic spices and fruits come from the Mediterranean area to Europe

1271—1202 the Venetian MARCO POLO travels through several Asian countries and tells of the culture and harvesting of typical herbs and spices

13th century snow is transported regularly from the Syrian and 'ebanese mountains to the Egyptian court at Cairo, where it is used for conserving and cooling special foods and meals

1300 smoke-houses exist in southern Sweden which are used for the smoking of fish

1342 the Arabian tradesman and physician IBN BATUTA mentions the expansion of the tea-drinking custom from China to India

1400 distillation methods are so far developed that high-percentage alcohol is available as an extraction solvent for aroma components in herbs and spices

1411 Venetian ships import spices with a value of 540000 ducats per year to Europe, where the consumption increases more and more. The cargoes come from Alexandria at the end of the 'spice road' which runs from South Arabia parallel to the coast of the Red Sea

1492—1503 Columbus on the newly discovered West Indian islands looks without success for the expected herbs and spices. He finds only a special variety of hot bell-pepper

1499 the Portuguese sailor and navigator Vasco da Gama returns from India with a load of pepper for the first time. In the following years Portuguese ships break the Arabian/Venetian trade monopoly for Asiatic spices

1500—1550 the preparation of a coffee beverage has expanded to Arabia. Coffee-houses in Cairo are known before 1500. After formation of the union between Turkey, Arabia and Egypt, coffee arrives in Europe first in Constantinople where 1534 public coffee-houses are known

the tomato, of Mexican origin, reaches Italy as a small yellow fruit. Because of the interesting aroma intensive breeding is started there. The general spread throughout Europe, however, is delayed. Several German papers of this time discuss the tomato as a 'poisoning plant'

in the well-known 'book of distillation' written by the Strasbourg doctor Brunschwig the production of essential oils and the distillation of spices and resins are described in detail

1502 Columbus finds the cocoa bean on an Amerindian trading boat near the Middle American coast, but does not recognize its flavour and importance

1516 a 'purity law' for beer is established in Germany, forbidding the use of additives other than malt, hops, yeast and water

1520 the Spanish conqueror Cortez describes an extensive culture of cocoa trees in Mexico, the use of cocoa beans as coins and the preparation of a cocoa beverage. For flavouring, the natives use a special spice previously unknown in Europe: black vanilla pods

1550 many recipes and formulae appear in the Italian and French literature describing the preparation of special sweets like marchpane, confectionery or candied fruits and spices. At the same time the consumption of liqueurs starts in Italy

1558 Gessner describes the flavouring of almonds by storage in the presence of violet flowers followed by squeezing of the flavourized oil. This idea is the basis of the *enfleurage* process widespread in southern France since the 17th century; odorous components volatilized from flowers are adsorbed on large surfaces of lard-impregnated paper

1570 the town council of Frankfurt takes possession of a mill for spice grinding in order to prevent adulteration

1580 the establishment of sugar refineries in Amsterdam begins. 80 years later there are 60 large factories in the Netherlands

1585 bell peppers, which originate in America, are cultivated in Spain and in Bohemia

a first load of chocolate produced in the monasteries of the New World arrives in Europe. To adapt the cocoa flavour to European taste two major contributions have been made: the use of vanilla and the cultivation of sugar-cane in America

1587 CAMERARIUS delivers his dissertation to the University of Marburg; he discusses the mechanism of odour perception, a classification of odours and the relationship between odour and taste

1598 the use of 108 distilled essential oils in medicine, food production and in the trade market can be proved from the *Pharmacopoeiae* of Nürnberg and Augsburg

16th century oranges come to Europe for the first time from their origin in China by way of India, Persia, Byzantium and the Mediterranean area

many books dealing with types and application of native aromatic herbs and spices are published in Germany

in Italy the production of ice-cream is simplified by addition of salpetre to the raw ice to induce lower temperatures

1600 sugared frozen fruit juices are transported from southern Italy and Spain to the courts of northern European countries

17th century at the courts of Spain, France, Italy and Austria chocolate is the preferred breakfast beverage. Cocoa products remain the province of the rich people for more than two centuries

1610 the first large loads of tea are transported from the Chinese island Formosa to the Netherlands by the East India Company

1625 the first coffee-house exists in Venice. The opening of such public coffee-house marks the expansion of coffee consumption in Europe. The first ones are opened in London (1652), Marseille (1659), Paris (1671), Hamburg (1677), Vienna (1683), Leipzig (1694), and Berlin (1721)

1636 a formula in a Chinese manuscript describes the production of rice wine after application of special types of malt and liquidifying

1630—1660 large loads of black tea arrive in Russia from China via Mongolia. At the same time, tea comes by sea to Holland and England. In these three countries tea consumption is introduced to all classes of the population by the end of the 17th century. Other countries follow the custom of tea drinking one or two centuries later

1665 at the court of the French King Louis XIV the luxury of sweet dishes reaches its height. The cook of Marshal Du Plessis-Pralin invents the *praliné*, which was originally an almond roasted in sugar

1670 the production of coffee in Arabia amounts to 40000 sacks per year. They are shipped to Mecca, Suez and Cairo

1671 the Florentine Procopio renders the ice-cream accessible to everyone; previously, it was a speciality for the courts only. Five years later, 250 masters of this profession produce the increasing amounts demanded by the people. In 1675 Procopio founds the first ice-cream shop in Paris which still exists as a café

1676 European merchants begin large tea imports from China after signing contracts

1708 Camembert cheese is mentioned in a French dictionary. The first French plant starts industrial production a century later (1813)

1721 a dictionary published in Leipzig mentions roasted grain as a substitute for coffee

1744 a 'wedding order' in Hanover states which classes are allowed to consume sugar-rich desserts at wedding breakfasts

1747 the Berlin chemist Marggraf finds sugar in the beet, and reports that on isolation it appeared 'pure and in high amounts, solid and dry, in nice hard crystals'

1756 the Swedish botanist Linnaeus classifies the cocoa bean under the name *Theobroma cacao* (= 'divine meal'). He tried to order odourous plants into seven odour classes

1762 coffee is now grown in large amounts in Brazil

1767 Priestley discovers that carbon dioxide induces a sour taste in water. The production of beverages impregnated with carbon dioxide is begun in Philadelphia in 1785

1770 Scheele isolates crystalline tartaric acid from grapes. In 1784 he finds an acid in lemons and calls it 'citric acid'

1775 French colonists and ships break the Dutch monopoly in the production of cloves. In previous decades this monopoly was guaranteed by the destruction of all cultures outwith certain Indonesian islands

widespread horticultural production of bell pepper is transmitted from Hungary. The plant came from Turkey with the conquering Turks via Bulgaria, where this plant was also cultivated very early

1780 the production of ice-cream is extended from France and Italy to the rest of Europe, now not only in soft but also in hard form as 'ice-tarte'

1782 Krünitz, owner of a coffee-roasting plant in Berlin, discusses the advantage of the filtration of coffee for acceptance and palatability

1799 Achard begins the production of sugar from beets in Germany after breeding for a high sugar content. The first industrial tests yield 6 % of the raw material

1804—1810 the French cook Appert develops procedures for ensuring unlimited stability of meat, vegetables, fruits and fruit juices by sterilization in closed vessels. This provides a basis for the worldwide spread even of flavour-rich food specialities which are liable to degradation

1811 the German pharmaceutical chemist Kirchhoff produces glucose from starch by means of dilute mineral acids

1817 the first elementary analysis of an essential oil is carried on in France by Houton de Labillardiére. He finds a ratio of five carbon atoms to eight hydrogen atoms, which is characteristic for terpenes

1821 Döbereiner develops the first rapid procedure for vinegar processing from wine. It is introduced few years later

1825 in his *Physiology du Gout* Brillat-Savarin describes gastronomy and taste physiology as it was seen in the early 19th century. This first modern standard work, written in a charming anecdotal form, remains valid for a century

the French physiologist Lassaigne concludes from experiments that the consumption of bread crust causes four times as much saliva production as the same amount of bread crumb. This is the first significant proof of the influence of the aroma intensity on digestibility

1826 Zeise and van Dyke first apply the steam distillation for the production of essential oils and find lighter and improved qualities

1833 Mitscherlich discovers nitrobenzene and describes its odour as 'between bitter almond and cinnamon'

1837 Liebig and Wöhler prove the formation of benzaldehyde, hydrocyanic acid and sugar by enzymatic splitting of amygdalin

1846 Liebig produces meat extract from 102 chickens and 300 pounds of meat. 20 years later the industrial production of 'Liebig's Meat Extract' is begun in South America

1847 Falconer reports the disappearance of the sweet sensation after chewing leaves of the Indian or African plant *Gymnema sylvestre*

1851 fruit esters are exhibited in London for the first time. Synthesized 'pear oil' (amyl acetate), 'apple oil' (amyl valerianate) and 'pineapple oil' (ethyl butyrate) are shown at the Great Exhibition

1852 DANIELL writes about the taste-modifying effect of berries of the Nigerian 'miracle fruit' (*Synsepalum dulcificum*). After chewing this, sour components are registered as sweet

1858 GOBLEY isolates vanillin from the vanilla pod

1859 in Great Britain the first patent for the production of fruit juices based on grapefruit is registered. Widespread consumption follows only a century later

1860 the development of the oil-refining industry makes available large amounts of petrol ether; this is soon used for the extraction of aroma components

1861 after many experiments BUTLEROW is successful in the total synthesis of a sugar by reacting alkali with formic aldehyde solution

1863 benzaldehyde is substituted for the poisonous nitrobenzene which had been introduced 30 years before as a flavouring with bitter almond character

1873 RASCHIG introduces a special type of separation column filled with granules. This equipment improves the distillation technique in such a way that essential oils can now be separated

1874 the chemical structure of vanillin from vanilla pods is determined by TIEMANN and HAARMANN

1875 after eight years of experimental work, PETER is successful in the development of a milk chocolate in Switzerland. This product wins a silver medal at the World Exhibition in Paris in 1878

1876 LINDE constructs compressive equipment using ammonia for refrigeration purposes. It now becomes possible to conserve aroma-rich foods in the fresh state for a long time

1878 FAHLBERG discovers the sweetening power of benzoic acid sulphimide, and registers a patent under the new name 'saccharin'

1884 in cooperation with the flavour industry which starts to develop in central Europe, WALLACH begins the systematic isolation, characterization and production of terpene components from essential oils

1886 FISCHER and PENZOLD determine threshold values of odorous components for the first time. They spray known amounts of ethyl mercaptan and chlorophenol into a room of known volume, and register the odour effect

1887 WALLACH formulates the 'isoprene rule' after recognition of the fact that most terpenes and terpenoides can be hypothetically constructed by joining isoprene units

1894 maltol is discovered in malt

1895 ZWAARDEMAKER classifies odours into 9 classes and 30 sub-classes

1900 aroma components which do not occur in nature are synthesized by the developing flavour industry

1901 at the Pan-American Exhibition a soluble coffee powder is shown. Unsolved aroma problems, however, delay its introduction for decades to come

1906 the Russian scientist TSWETT publishes his classical experiment on the separation of coloured compounds of leaves by means of sorption processes. This principle will be the basis of modern chromatographic techniques used half a century later for separating aroma components

1909 a special decree in France limits the district where the brandy produced is permitted to be labelled as 'cognac'

1910 the identification of taste and flavour enhancers in typical East Asian meals begins in Japan. In subsequent years monosodium glutamate is found in seaweed and inosinic acid in the Japanese Bonito mushroom. Only half a century later, however, is the industrial production of such substances realized

1912 the French scientist MAILLARD studies the non-enzymatic browning reaction between sugars and protein degradation products which is characterized later as the MAILLARD reaction

1918 systematic breeding of bell pepper is started in Hungary. Soon aroma-rich vegetable varieties and hot varieties can be distingguished

1920 POWERS and CHESNUT register the emanation of acetaldehyde from ripe apples

RUZICKA in Switzerland starts systematic studies in the terpene field, elucidating the different structures in subsequent years. In 1928 muscone and zibetone are synthesized

1925 STAUDINGER and REICHSTEIN begin research on an isolated coffee oil. Using classical techniques only, they find and tabulate several dozen aroma compounds

1937 AUDRIETH and SVEDA discover the suitability of cyclamate as a sweetener. Large-scale industrial production will not begin before 1950

1944 new processes for aroma recovery in the course of the technical fruit juice concentration are developed in the USA

1950 not more than six aroma components are known in bread. 20 years later more than 200 volatiles have been identified

1952 AMOORE's 'stereochemical theory' of odour perception proposes that the olfactory receptors are sensitive to the size, shape and electronic status of the odorant

a routine control for toxicological effects results in the prohibition of one of the most important aroma components, cumarine

1955 a systematic elucidation of food aroma spectra by means of chromatographic methods begins. 25 years later, 4500 aroma components are known

1958 the Australian scientists HERCUS and MORISSON first identify aroma compounds by their mass spectrum

1960 BUTENANDT isolates and identifies the pheromone (sex attractant) of the silkworm moth *Bombyx mori* as hexadeca-10-trans, 12-cis-dien-1-ol

1961 BUTTERY and TERANISHI use head-space technique for the first time in studies on fruit juices and initiate a new phase in aroma analysis

1967 SHALLENBERGER and ACREE publish the first modern reception theory for the sweet taste, postulating the presence of two electronegative atoms with one hydrogen atom at a defined distance

1968 in Great Britain HARPER et al. summarize and define 44 terms for food odours. Ten years later this list is enlarged to more than 100 descriptors

1970 the worldwide success of the new sweetener cyclamate is stopped by the results of new animal tests which are, however, controversial

1975 only 1% of the world vanillin production comes from vanilla pods, compared with 80% from the splitting of lignin

from all flavouring additives 75% are of natural origin; only 1% is synthetic and not 'natur-identical'

a new task for aroma research begins, with the flavouring of new simulated or dietetic foods which are neutral in flavour

1978 the production of flavour enhancers amounts to 66 million pounds in the USA. For 1990 a requirement of 290 million pounds is estimated

1982 AMOORE defines the composition of flavourings by combination of single components as 'an artistic work like great music or painting'.

Literature

ACREE, T. E. and J. BARNARD, The analysis of odor-active volatiles in gas chromotographic effluents, in P. Schreier, Analysis of volatiles, W. de Gruyter Berlin/New York 1984, p. 251-267

ADACHI, A., Y. KAWAMURA, M. OHARA and S. IKEDA, Neurophysiological studies on taste effectiveness of sodium-5′-guanylate as a chemical taste enhancer, Symp. Amino Acid and Nucleic Acid **12**, 63—68 (1965)

ADAM, S., Sorptive enrichment and analysis of volatile compounds in drinking juices, in P. SCHREIER, Analysis of volatiles, W. de Gruyter Berlin/New York 1984, p. 419—432

ÅKESSON, C., T. PERSSON and E. VON SYDOW, Predicting sensory qualities and preference values of aroma and flavour by use of gas chromatographic data, Work Doc. IV. Internat. Congress Food Sci. and Technol. Madrid 1974, theme 3, p. 22—24

ALBRECHT, A. A. and L. M. BIEDLER, Flavour mystery lessens through R & D, Food Engng. **38**, 102—104 (1966)

AMERINE, M. A., R. M. PANGBORN and E. B. ROESSLER, Principles of sensory evaluation of food, Acad. Press, New York 1965, p. 275

AMOORE, J. E., Odor theory and odor classification, in E. T. THEIMER, Fragrance chemistry. The science of the sense of smell, Acad. Press New York 1982, p. 73

AMOORE, J. E., The stereochemical specifities of human olfactory receptors, Parfum. essent. Oil Rec. **43**, 321—323 (1952)

AMOORE, J. E., Progress towards some direct quantitative comparisons of the stereochemical and vibrational theories of odor, in G. OHLOFF and A. F. THOMAS, Gustation and olfaction, Acad. Press London, New York 1971, p. 147—164

AMOORE, J. E. and D. VENSTROM, Sensory analysis of odor qualities in terms of the stereochemical theory, J. Food Sci. **31**, 118—128 (1966)

ANDERSON, D. F. and E. A. DAY, Quantitative evaluation and effect of certain microorganisms on flavor components of Blue cheese, J. agric. and Food Chem. **14**, 241—245 (1966)

ANGELINE, J. F. and G. F. LEONARDOS, Food additives. Some economic considerations, Food Technol. **27**, 4, 40—50 (1973)

Anon., Süßstoffe — Kurzreferate und Podiumsdiskussion, Symposium calorienarme u. calorienfreie Lebensmittel, Ernährungs-Umschau **17**, 14—18 (1970)

Anon., Definition of the words 'flavour' and 'a flavour', Food Technol. Australia **22**, 328 (1970)

ARNOLD, R. G., K. M. SHAHANI and B. K. DWIVEDI, Application of lipolytic enzymes to flavor development in dairy products, J. Dairy Sci. **58**, 1127 1143 (1975)

ASKAR, A. and H. J. BIELIG, Geschmacksverbesserung von Lebensmitteln. I. Zusatz von geschmacksaktiven Verbindungen, Alimenta **15**, 3—13 (1976)

BADINGS, H. T. and C. de JONG, Headspace analysis for the study of aroma compounds in milk and dairy products, in P. SCHREIER, Analysis of volatiles, W. de Gruyter Berlin/New York 1984, p. 401—417

BAKER, D., M. H. NEUSTADT and L. ZELENY, Relationship between fat acidity values and types of damage in grain, Cereal Chem. **36**, 308—311 (1959)

Baldry, J., Flavour volatiles of tetraploid banana fruit, Fruits **37**, 699—704 (1982)

BARTOSHUK, L. M., C. H. LEE and R. SCARPELLINO, Sweet taste of water induced by artichoke (*Cynara scolymus*), Science **178**, 988—989 (1972)

BARYLKO-PIKIELNA, N., M. ROTHE, E. PIETRZAK and A. JACKOWSKA, Sensory characteristics of 2-heptanone. Part I. Threshold values, concentration/odour intensity relations, methodological aspects, Nahrung **27**, 391—401 (1983)

BEAUCHAMPS, G. K. and M. MORAN, Dietary experience and sweet taste preferences in human infants, Appetite **3**, 139—152 (1982)

BEDNARCZYK, A. A. and A. KRAMER, Statistical preselection of significant flavor contributors speeds development of synthetic ginger flavor and flavor-fortified ginger oil, Food Technol. **25**, 1098—1107 (1971)

BEETS, M. G. J., Odor and stimulant structure, in E. T. THEIMER, Fragrance chemistry. The science of the sense of smell, Acad. Press New York 1982, p. 77—122

BEMELMANS, J. M. H., Isolation and concentration of volatiles from foods. Isolation and concentration from the product phase, in H. MAARSE and R. BELZ, Isolation, separation and identification of volatile compounds in aroma research, Akademie-Verlag Berlin 1981, p. 4—37

BEMELMANS, J. M. H. and M. C. TEN NOEVER DE BRAUW, Analysis of off-flavours in food, Proceed. Internat. Symposium Aroma Research Zeist/Netherlands 1975, Centre Agric. Publish. and Doc. (Pudoc) Wageningen 1975, p. 85—93

BERRY, R. E., P. E. SHAW, J. H. TATUM and C. W. WILSON III, Citrus oil flavour and composition studies, Food Technol. **37**, 12, 88—91 (1983)

BERTINO, M., G. K. BEAUCHAMPS and K. ENGELMAN, Long-term reduction in dietary sodium alters the taste of salt, Amer. J. Clin. Nutrit. **36**, 1134—1144 (1982)

BERTINO, M., G. K. BEAUCHAMPS and K. C. JEN, Rated taste perception in two cultural groups, Chem. Senses and Flavour **8**, 3—15 (1983)

BEYELER, M. and J. SOLMS, Interaction of flavor model compounds with soy protein and bovine serum albumin, Lebensmittel-Wiss. u. Technol. **7**, 217—219 (1974)

BIGGERS, R. E., J. J. HILTON and M. A. GIANTURCO, Differentiation between coffea arabica and coffea robusta by computer evaluation of gas chromatographic profiles. Comparison of numerically derived quality predictions with organoleptic evaluations, J. chromatogr. Sci. **7**, 453—464 (1969)

BOECKH, J., Neurophysiologische Grundlagen des Geruchssinns, in F. DRAWERT, Geruch- und Geschmackstoffe. Internat. Symposium Fa. Haarmann u. Reimer, Bad Pyrmont 1974, Verlag Hans Carl Nürnberg 1975, p. 1—18

BOELENS, H., P. J. DE VALOIS, D. DE RIJKE, H. J. WOBBEN and H. J. TAKKEN, Alternative methods in flavour research, Work Doc. IV th Internat. Congress Food Sci. and Technol., Madrid 1974, 1 a, p. 28—30

BOGGS, M. B., R. G. BUTTERY, D. W. VENSTROM and M. L. BELOTE, Relation of hexanal in vapor above stored potatoe granules to subjective flavor estimates, J. Food Sci. **29**, 487—489 (1964)

BOLAND, W., P. NEY, L. JAENICKE and G. GASSMANN, A (closed-loop stripping) technique as a versatile tool for metabolic studies of volatiles, in P. SCHREIER, Analysis of volatiles, W. de Gruyter Verlag Berlin/New York 1984, p. 371—380

BOMBEN, J. L., Aroma recovery and retention in concentration and drying of foods, Advances Food Res. **20**, 1—111 (1973)

BOMBEN, J. L., D. G. GUADAGNI and J. G. HARRIS, Aroma concentration for dehydrated foods, Food Technol. **23**, 83—86 (1969)

BRODY, S. S. and J. CHANEY, Flame photometric detector, J. Gas Chromatogr. **4**, 42—46 (1966)

BRUEMMER, J. H., Aroma substances of citrus fruits and their biogenesis, in F. DRAWERT, Geruch- und Geschmackstoffe, Internat Symposium Fa. Haarmann u. Reimer, Bad Pyrmont 1974, Verlag Hans Carl Nürnberg 1975, p. 167—176

BUSER, H.-R., C. ZANIER and H. TANNER, Identification of 2,4,6-trichloroanisol as a potent compound causing cork taint in wine, J. agric. and Food Chem. **30**, 359—362 (1982)

BUTTERY, R. G., D. G GUADAGNI and J. HARRIS, Odour intensities of hop oil compounds, J. Sci. Food and Agric. **17**, 142—144 (1966)

BUTTERY, R. G., D. G. GUADAGNI and L. C. LING, Flavour compounds: volatilities in vegetable oil and oil-water mixtures. Estimation of odor thresholds, J. agric. and Food Chem. **21**, 198—201 (1973)

BUTTERY, R. G., D. G. GUADAGNI and R. E. LUNDIN, Some 4,5-dialkyl thiazols with potent bell pepper-like aromas, J. agric. and Food Chem. **24**, 1—3 (1976)

BUTTERY, R. G., W. F. HADDON, R. M. SEIFERT and J. G. TURNBAUGH, Thiamin odor and bis (2-methyl-3-furyl) disulfide, J. agric. and Food Chem. **32**, 674—676 (1984)

BUTTERY, R. G., R. M. SEIFERT, D. G. GUADAGNI, D. R. BLACK and L. C. LING, Characterization of some volatile constituents of carrots, J. agric. and Food Chem. **16**, 1009—1015 (1968)

BUTTERY, R. G. and R. TERANISHI, Gas-liquid chromatography of vegetables and fruits. Direct injection of aqueous vapors, Anal. Chem. **33**, 1439—1441 (1961)

CAIN, W. S., Odour intensity: differences in the exponent of the psychophysical functions, Percept. Psychophys. **6**, 349—354 (1969)

CAIRNCROSS, S. E. and L. B. SJÖSTRÖM, Flavour profiles — a new approach to flavour problems, Food Technol. **4**, 308—311 (1950)

CASIMIR, D. J. and F. B. WHITFIELD, Flavour impact values: a new concept for assigning numerical values for the potency of individual flavour components and their contribution to the overall flavour profile, Report Symposium Flavours Fruits and Fruit Juices, Bern 1978, Internat. Federat. Fruit Juices' Producers, Sci.-techn. Comm. no. 15, p. 325—347

CAUL, J. F., The profile method of flavour analysis, Advances Food Res. **VII**, 1—40 (1957)

CAUL, J. F., Food flavour and the consumer, Ber. 5. Welt-Getreide- u. Brotkongr. Dresden 1970, VEB Fachbuchverlag Leipzig 1970, 5, p. 247—251

CHANG, S. S., Overcoming problems in flavour component identification, Food Technol. **27**, 4, 27—39 (1973)

CHRISTENSEN, C. M. and Z. M. VICKERS, Relationships of chewing sounds to judgements of food crispness, J. Food Sci. **46**, 574—578 (1981)

CHRISTENSEN, C. M., Effects of color on aroma, flavour and texture judgements of foods, J. Food Sci. **48**, 787—790 (1983)

CLARK, R. G. and D. A. CRONIN, The use of activated charcoal for the concentration and analysis of headspace vapours containing food aroma volatiles, J. Sci. Food and Agric. **26**, 1615—1624 (1975)

CLARK, R. G. and H. E. NURSTEN, Analysis of the sensory term 'nutty' and a list of compounds claimed to be nutty, Internat. Flavours and Food Additives **8**, 197—201 (1977)

CLONINGER, M. R. and R. E. BALDWIN, L-Aspartyl-L-phenylalanine methylester (Aspartame) as a sweetener, J. Food Sci. **39**, 347—349 (1974)

CONSDEN, R., A. H. GORDON and A. J. P. MARTIN, Qualitative analysis of proteins: partition chromatographic method using paper, Biochem. J. **38**, 224—232 (1944)

COOMES, T. J., British and international legislative

control of flavouring substances in food, Proc. Internat. Symposium on Aroma Research Zeist/Netherlands, Centre for Agricultural Publish. and Doc. (Pudoc) Wageningen 1975, p. 197—202

CROTEAU, R., Biosynthesis of monoterpenes and sesquiterpenes, in F. DRAWERT, Geruch- und Geschmackstoffe, Internat. Symposium Fa. Haarmann & Reimer, Bad Pyrmont 1974, Verlag Hans Carl Nürnberg 1975, p. 153—166

CUNNINGHAM, D. G., T. E. ACREE, J. BARNARD, R. M. BUTTS and P. A. BRAELL, Charm analysis of volatiles, Food Chem. 19, 137—147 (1986)

CURTIS, R. F., C. DENNIS, J. M. GEE, M. G. GEE, N. M. GRIFFITHS, D. G. LAND, J. L. PEEL and D. ROBINSON, Chloroanisoles as a cause of musty taint in chicken and their microbiological formation from chlorophenols in boiler house litter, J. Sci. Food and Agric. 25, 811—828 (1974)

DAENIKER, H. U., Lebensmittelaromen. Bedeutung, Herstellung und Anwendung, Neue Züricher Zeitung 1975, no. 29, 45—46

DANIEWSKI, M. and Z. MIELNICZUK, Study into the applicability of isolation methods and injection into a chromatographic column of volatile products of Maillard's reaction in model systems, Acta aliment. Acad. Sci. hung. 3, 195—206 (1974)

DASTOLI, F. R., V. D. LOPIEKES and A. R. DOIG, Bitter-sensitive protein from porcine taste buds, Nature 218, 884—885 (1968)

DASTOLI, F. R. and S. PRICE, Sweet-sensitive protein from bovine taste buds: isolation and assay, Science 154, 905—907 (1966)

DAWSON, E. H., J. L. BROGDON and S. MCMANUS, Sensory testing differences in taste. 2. Selection of panel members, Food Technol. 17, 10, 39—44 (1963)

DAY, E. A., D. A. LILLARD and M. W. MONTGOMERY, Autoxidation of milk lipids. III. Effect on flavour of the additive interactions of carbonyl compounds at subthreshold concentrations, J. Dairy Sci. 46, 291—301 (1963)

DESOR, J. A., Sweet: the special taste, Cereal foods world 21, 69—70, 88 (1976)

DESOR, J. A. and L. S. GREENE, Preferences for sweet and salty in 9- to 15-year-old and adult humans, Science 190, 686—687 (1975)

DHONT, J. H., Vapour pre-adsorption thin-layer chromatography, J. Chromatogr. 90, 203—207 (1974)

DHONT, J. H., Thin-layer chromatography, in H. MAARSE and R. BELZ, Handbuch der Aromaforschung: Isolation, separation and identification of volatile compounds in aroma research, Akademie-Verlag Berlin 1981, p. 100—152

DHONT, J. H., C. VINKENBORG, H. COMPAAN, F. J. RITTER, R. P. LABEDIE, A. VERWEIJ and R. A. DE ZEEUW, Application of R_F correction in thin-layer chromatography by means of two reference R_F values, J. Chromatogr. 47, 376—381 (1970); 71, 283—289 (1972)

DHONT, J. H. and C. WEURMAN, The isolation and determination of volatile compounds by adsorption on charcoal, Analyst 85, 419—422 (1960)

DIJKSTRA, G. and J. DE GOEY, The use of coated capillaries instead of columns for gas chromatography, in D. H. DESTY, Gas Chromatography 1958, Butterworths London 1958, p. 56—62

DINSMORE, H. L. and S. NAGY, Colorimetric furfural measurement as an index of deterioration in stored citrus juice, J. Food Sci. 37, 768—770 (1972)

DITTRICH, H. H. and E. KERNER, Diacetyl als Weinfehler. Ursache und Beseitigung, Wein-Wiss. 19, 528—535 (1964)

VAN DEN DOOL, H., Aromatization and international legislation, Proc. Internat. Symposium on Aroma Research Zeist/Netherlands 1975, Centre for Agric. Publish. and Doc. (Pudoc) Wageningen 1975, p. 239—242

DOVING, K. B., Problems in the physiology of olfaction, in H. W. SCHULTZ, E. A. DAY and L. M. LIBBEY, Symposium on Foods: The Chemistry and Physiology of Flavours, The Avi Publish. Co. Westport/Conn. 1967, p. 80

DRAKE, B. and B. JOHANSSON, Sensory evaluation of food, 1 and 2, Swed. Inst. Food Preservat. Göteborg, SIK-Rapport no. 255 (1969)

DRAVNIEKS, A., Evaluation of human body odors: methods and interpretations, J. Soc. Cosmet. Chem. 26, 551—571 (1975)

DRAWERT, F., Biochemical formation of aroma components, Proc. Internat. Symposium on Aroma Research Zeist/Netherlands 1975, Centre Agric. Publish. and Doc. (Pudoc) Wagening 1975, p. 13—39

DRAWERT, F. and R. BERGER, Possibilities of the biotechnical production of aroma substances by plant tissue cultures, in P. SCHREIER, Flavour '81, W. de Gruyter Berlin/New York 1981, p. 509—527

DRAWERT, F. and R. F. BERGER, Über die Biogenese von Aromastoffen bei Pflanzen und Früchten. XIX. Mitt.: Vergleich der Biosynthese von Aromastoffen in Segment-, Kallus- und Suspensionskulturen von Citrusarten, Chemie, Mikrobiol., Technol. Lebensmittel 7, 143—147 (1982)

DRAWERT, F. and R. F. BERGER, Über die Biogenese von Aromastoffen bei Pflanzen und Früchten. XX. Mitt.: Abhängigkeit der Aromabiosynthese in Erdbeerfrüchten von exogenen Parametern, Lebensmittel-Wiss. u. -Technol. 16, 209—214 (1983)

DRAWERT, F. and N. CHRISTOPH, Significance of the sniffing technique for the determination of odour thresholds and detection of aroma impacts of trace volatiles, in P. SCHREIER, Analysis of volatiles, W. de Gruyter Berlin/New York 1984, p. 269—291

DRAWERT, F., W. HEIMANN, R. EMBERGER and R. TRESSL, Über die Biogenese von Aromastoffen bei Pflanzen und Früchten. II. Enzymatische Bildung von Hexen-(2)-al-(1), Hexanal und deren Vorstufen, Liebigs Ann. Chemie 693, 200—208 (1966)

DUERR, P. and SCHOBINGER, Möglichkeiten zur Beurteilung von Aromakonzentraten in der Praxis, in Ber. Internat. Fruchtsaft-Union, Symposium Neue Erkenntnisse auf dem Gebiet der Frucht- und Gemüsesäfte, Juris-Verlag Zürich 16, 235—241 (1980)

DUMONT, J. P., S. ROGER, P. Cerf and J. ADDA, Etude de composes volatiles neutres presents dans le Vacherin, Lait **54**, 243—251 (1974)

DWIVEDI, B. K. and J. E. KINSELLA, Continuous production of Blue-type cheese flavor by submerged fermentation of Penicillium roqueforti, J. Food Sci. **39**, 620—622 (1974)

VAN EIJK, A., Flavourings in deposited articles. Adaptation of flavouring substances to the raw material and to the end-product, Flavour-Ind. **2**, 347—352 (1971)

EISENBRAND, J., Über das Molekulargewicht der Aromastoffe, Mitt.-Bl. GDCh-Fachgruppe Lebensmittelchemie u. gerichtl. Chemie **25**, 281—284 (1971)

EISERLE, R. J. and W. J. DOWNEY, A review of the literature concerned with flavor research as it applies to the problems of the flavor industry, Crit. Rev. in Food Technol. **2**, 159 (1971)

EL-DASH, A. A. and J. A. JOHNSON, Protease enzymes: effect on bread flavor, Cereal Sci. today **12**, 282 to 288 (1967)

ELLIS, G. P., The Maillard reaction, Advances Carbohydrate Chem. and Biochem. **14**, 63—134 (1959)

EMCH, F., Aromakonzentration und Aromagewinnung, in J. SOLMS and H. NEUKOM, Aroma- und Geschmacksstoffe in Lebensmitteln, Forster-Verlag Zürich 1967, p. 237—252

ENGEL, C., A. P. DE GROOT and C. WEURMAN, Tetrachloranisol: a source of musty taste in eggs and broilers, Science **154**, 270—271 (1966)

ENGEL, K.-H. and R. TRESSL, Formation of aroma components from nonvolatile precursors in passion fruit, J. agric. and Food Chem. **31**, 998—1002 (1983)

ERIKSSON, C., Aroma compounds derived from oxidized lipids. Some biochemical and analytical aspects, J. agric. and Food Chem. **23**, 126—128 (1975)

FAZZALARI, F. A., Compilation of odour and taste threshold values data, Amer. Soc. Test. and Materials DS 48A, Philadelphia 1978

FERON, R. and M. GOVIGNON, The relation between free acidity and taste of edible oils, Ann. Falsificat. et Expertise chim. **54**, 308—311 (1961)

FINCKE, H, Über Aromastoffe unter besonderer Berücksichtigung ihrer Bedeutung für die Ernährung und ihrer Verwendung im Lebensmittelgewerbe und bei der Speisenzubereitung, Z. Lebensmittel-Untersuch. u. -Forsch. **89**, 1—18 (1949)

FISCHER, R., Gustatory, behavioral and pharmacological manifestations of chemoreception in man, in G. OHLOFF and A. F. THOMAS, Gustation and Olfaction, Academic Press London, New York 1971, p. 220

FLANNARY, R. J., Nonagricultural sources of food, Food Technol. **29**, 8, 64—66 (1975)

FLATH, R. A., C. R. BLACK, D. G. GUADAGNI, W. H. MCFADDEN and T. H. SCHULTZ, Identification and organoleptic evaluation of compounds in Delicious apple essence, J. agric. and Food Chem. **15**, 29—35 (1967)

FLINK, J. M., Mechanisms of retention of organic volatiles in freeze-dried systems, J. Food Technol. **7**, 199—211 (1972)

FRANCIS, F. J., Colour and appearance as dominating sensory properties of foods, in G. G. BIRCH, J. G. BRENNAN and K. J. PARKER, Sensory properties of foods, Appl. Sci. Publish. London 1977. p. 27—43

FRANZKE, C., Bemerkungen über die jodometrischen Verfahren zur Bestimmung der Peroxydzahl in Nahrungsfetten, Z. Lebensmittel-Untersuch. u. -Forsch. **103**, 108—112 (1956)

FRICKER, A., Sensorische Aspekte der Lebensmittelverarbeitung, Internat. Z. Vitamin- u. Ernähr. -Forsch., Beiheft Nr. 14 "Qualitätskriterien der Nahrung", **1974**, p. 179—190

GALANOS, D. S. and V. M. KAPOULAS, The paper chromatographie identification of compounds using two reference compounds, J. Chromatogr. **13**, 128—138 (1964)

GARBUSOV, V., G. REHFELD, G. WÖLM, R. V. GOLOVNJA and M. ROTHE, Volatile sulfur compounds contributing to meat flavour. I. Components identified in boiled meat, Nahrung **20**, 235—241 (1976)

VAN GEMERT, L. J., Coordination of sensory and instrumental analysis, in H. MAARSE and R. BELZ, Isolation, separation and identification of volatile compounds in aroma research, Akademie-Verlag Berlin 1981, p. 240—258

GERVAIS, P. and F. SAUVAGEOT, Un exemple de batterie d'epreuves permettant d'évaluer les aptitudes sensorielles individuelles pour la dégustation de denrées alimentaires, Lebensmittel-Wiss. u. Technol. **17**, 60—73 (1984)

GIANTURCO, M. A., R. E. BIGGERS and B. H. RIDLING, Seasonal variations in the composition of the volatile constituents of black tea. A numerical approach to the correlation between composition and quality of tea aroma, J. agric. and Food Chem. **22**, 758—764 (1974)

GILLETTE, M., Applications of descriptive analysis, J. Food Protect. **47**, 403—409 (1984)

GIOCOMETTI, T., Free and bound glutamate in natural products. Glutamic acid, in L. J. FILER, Advances in Biochemistry and Physiology, Raven Press New York 1979, p. 25

GIRARDOT, N, F., D. R. PERYAM and R. SHAPIRO, Selection of sensory testing panels, Food Technol. **6**, 140—145 (1952)

GLATZEL, H., Physiologische Bedeutung der Würzstoffe, Dtsch. Lebensmittel-Rdsch. **64**, 1—5 (1968)

GLATZEL, H., Die Bedeutung der Riech- und Schmeckstoffe für Nahrungswahl und Bekömmlichkeit, Qualitas Plantarum et Mater. vegetabiles **15**, 280 to 288 (1968)

GÖTTINGER, E., K. HAGMÜLLER, M. HELLAUER und H. VINAZZER, Über die Wirksamkeit des Süßstoffes Zyklamat auf Leberparenchym, korrespondierende Fermente und Gerinnungsfaktoren des Blutes, Wiener klin. Wochenschr. **80**, 328—331 (1968)

GOLAY, M. J. E., Theory of chromatography in open and coated tubular columns with round and rectangular cross-sections, in D. H. DESTY, Gas Chromatography 1958, Butterworths London 1958, p. 36—53

GOLAY, M. J. E., Brief report on gaschromatographic theory, in R. P. W. SCOTT, Gaschromatography 1960, Butterworths Washington 1960, p. 139—143

GOLOVNJA, R. V., Investigation of organic bases in the specific odour of casein and coprecipitate during storage, Nahrung **26**, 603—613 (1982)

GOLOVNJA, R. V., V. N. SYOMIŇA, N. G. JAKOVLEVA and N. G. ENIKEEVA, List of chemical substances and uniform procedure for selection of panelists by their ability to recognize odours, Nahrung **30**, 111—118 (1986)

GOLOVNJA, R. V. and V. URALETS, Correlation isothermal index for temperature programmed gas chromatography, J. Chromatogr. **36**, 276—282 (1968)

GOLOVNJA, R. V. and V. URALETS, Gas chromatographic analysis of flavour components with correlation isothermal retention indices, J. Chromatogr. **61**, 65—71 (1971)

GOSTECNIK, G. F. and A. ZLATKIS, Computer evaluation of gas chromatographic profiles for the correlation of quality differences in cold pressed orange oils, J. Chromatogr. **106**, 73—81 (1975)

GREMLI, H. A., Interaction of flavour compounds with soy protein, J. amer. Oil Chemist's Soc. **51**, 95A—97A (1974)

GREMLI, H. and J. WILD, Enzymatic flavour regeneration in processed foods, Work Doc. IV. Internat. Congr. Food Sci. and Technol. Madrid 1974, 1a, p. 62—63

GRIFFITHS, N. M. and D. G. LAND, 6-chloro-o-cresol taint in biscuits, Chem. and Ind. [London] **1973**, 904

GRINKER, J. A., Taste factors and obesity, Food Technol, **29**, 8, 76—78 (1975)

GROB, K., Heutige Grenzen der hochauflösenden Gas-Chromatographie, Chromatographia **7**, 94—98 (1974)

GROEPLER, P. and W. SCHÄFER, Die Bestimmung der Fettsäurezahl als Nachweis der Gesundheit bei deutschem Weizen, Getreide u. Mehl **8**, 28—30 (1958)

GROSCH, W., G. LASKAWY and K. H. FISCHER, Aroma compounds formed by enzymic co-oxidation, Proc. Internat. Symposium on Aroma Research Zeist 1975, Centre Agric. Publish. and Doc. (Pudoc) Wageningen 1975, p. 75—83

GRUNDSCHOBER, F., Die internationale Organisation der Aromenindustrie, in W. AUERSWALD und B. M. BRANDSTETTER, Natürliche, naturidentische und synthetische Geschmacksstoffe und deren lebensmittelrechtliche Aspekte, Verlag für medizinische Wiss. Wilhelm Maudrich Wien 1973, p. 11—12, 28, 29, 31, 38

GUADAGNI, D. G., R. G. BUTTERY and J. HARRIS, Odor intensities of hop oil compounds, J. Sci. Food and Agric. **17**, 142—144 (1966)

GUADAGNI, D. G., R. G. BUTTERY and S. OKANO, Odour thresholds of some organic compounds associated with food flavours, J. Sci. Food and Agric. **14**, 761—765 (1963)

GUADAGNI, D. G., R. G. BUTTERY, S. OKANO and H. K. BURR, Additive effect of sub-threshold concentrations of some organic compounds associated with food aromas, Nature [London] **200**, 1288 to 1289 (1963)

GUADAGNI, D. G., R. G. BUTTERY and J. G. TURNBAUGH, Odour thresholds and similarity ratings of some potato chip components, J. Sci. Food and Agric. **23**, 1435—1444 (1972)

GUADAGNI, D. G. and I. C. MIERS, Statistical relationship between methyl sulfide content and aroma intensity in canned tomato juice, Food Technol. **23**, 375—377 (1969)

GUADAGNI, D. G., S. OKANO, R. G. BUTTERY and H. K. BURR, Correlation of sensory and gas-liquid chromatographic measurements of apple volatiles, Food Technol. **20**, 518—521 (1966)

GUBLER, B. A., H. A. GREMLI, G. VERDE and J. WILD, Some aspects of interactions between food and flavour components, Work Doc. IV. Internat. Congr. Food Sci. and Technol. Madrid 1974, 1d, 4, p. 41

HALL, R. L., Food flavors. Benefits and problems, Food Technol. **22**, 1388—1392 (1968)

HALL, R. L. and E. J. MERWIN, The role of flavors, Food Technol. **35**, 6, 46—51 (1981)

HALL, R. L. and B. L. OSER, Recent progress in the consideration of flavoring ingredients under the Food Additives Amendment. III. Gras substances, Food Technol. **19**, 253—256 (1965)

HARLEY, J., W. NEL and V. PRETORIUS, Flame ionization detector for gas chromatography, Nature [London] **181**, 177—178 (1958)

HARPER, R., Techniques of analysis of flavours. Sensory methods, in I. D. MORTON and A. J. MACLEOD, Food flavours, Part A. Introduction, Elsevier Sci. Publish. Co., Amsterdam, Oxford, New York 1982, p. 100

HARPER, R., E. C. BATE-SMITH, D. G. LAND and N. M. GRIFFITHS, A glossary of odour stimuli and their qualities, Perf. Essent. Oil Rec. **59**, 22—37 (1968)

HARRIS, W. E. and H. W. HABGOOD, Programmed temperature gas chromatography, Verlag Wiley New York 1966

HARRISON, G. A. F. and E. COLLINS, Determination of taste thresholds for a wide range of volatile and non-volatile compounds in beer, Proc ASBC Congress 1968, A. Guinness Dublin 1968, p. 83—87

HAUKELI, A. D., T. JACOBSEN and S. LIE, Regression analysis of some beer volatiles, Techn. quart. Master Brewers Assoc. Amer. **10**, 47—52 (1973)

HAYASE, F., T.-Y. CHUNG and H. KATO, Changes of volatile components of tomato fruit during ripening, Food Chem. **14**, 113—124 (1984)

HEROUS, G. R. and J. D. MORRISON, Mass spectrometric analyses. Appendix I to A. R. THOMPSON: Volatile products of apples. I. Identification of acids and alcohols, Austral. J. biol. Sci. B—4, 290—292 (1951)

HERRMANN, J., Neue Methoden der Be- und Auswertung sensorischer Eigenschaften von Lebensmitteln und der Berechnung ihrer Veränderungen. 2. Mitt. Die Berechnung sensorischer Veränderungen der Lebensmittel bei thermischen Prozessen aus reaktionskinetischen Daten, Nahrung **15**, 827—836 (1971)

HEWITT, E. J., D. A. M. MACKAY, K. KONINGSBACHER and T. HASSELSTROM, The role of enzymes in food flavors, Food Technol. **10**, 487—489 (1956)

HEYNS, K., R. STURE and H. PAULSEN, Bräunungsreaktionen und Fragmentierungen von Kohlenhydraten. Teil 1. Die flüchtigen Abbauprodukte der Pyrolyse von D-Glucose, Carbohydrate Res. **2**, 132 · 149 (1966)

HIRVI, T., Mass fragmentographic and sensory analyses in the evaluation of the aroma of some strawberry varieties, Lebensmittel-Wiss. u. -technol. **16**, 157—161 (1983)

HODGE, J. E., Origin of flavour in foods, Nonenzymatic browning reactions, in H. W. SCHULTZ, E. A. DAY and L. M. LIBBEY, Symposium on Foods: The Chemistry and Physiology of Flavors, The Avi Publish. Co. Westport/Conn. 1967, p. 465—491

HOPPE, K., Beiträge zur Psychophysik des Süßgeschmacks. 4. Mitt. Methodischer Nachtrag zum Einfluß der Versuchsanordnung und des Spülregimes bei Differenzprüfungen, Nahrung **28**, 983—993 (1984)

HUCKLE, M. T., A low temperature zone melting technique for the concentration of flavours, Chem. and Ind. **1966**, 1490—1492

HULPKE, H., Einige physiologische und biochemische Aspekte des "Off-Flavour"-Problems unter besonderer Berücksichtigung der Pestizideinwirkung auf Nahrungspflanzen, Qualitas plantarum et mater. vegetabiles **18**, 116—132 (1969)

IFT Sensory Evaluation Division, Sensory evaluation guide for testing food and beverage products, Food Technol. **35**, 11, 50—59 (1981)

INGLETT, G. E., B. DOWLING, J. J. ALBRECHT and F. A. HOGLAN, Taste-modifying properties of miracle fruit (Synsepalum dulcificum), J. agric. and Food Chem. **13**, 284—287 (1965)

ISSENBERG, P., Mass spectrometry for flavor research, Food Technol. **23**, 11, 103—110 (1969)

ISSENBERG, P., A. KOBAYASHI and T. J. MYSLIWY, Combined gas chromatography — mass spectrometry in flavor research, J. agric. and Food Chem. **17**, 1377—1386 (1969)

JAMES, A. T. and A. J. P. MARTIN, Gas-liquid partition chromatography: The separation and micro-estimation of volatile fatty acids from formic acid to dodecanoic acid, Biochem. J. **50**, 679—690 (1952)

JANAK, J., Gas chromatography as a sampling procedure for thin layer or paper chromatography, J. Gas Chromatogr. **1**, 10, 20—23 (1963)

JELLINEK, G. Introduction to and critical review of modern methods of sensory analysis (odour, taste and flavour evaluation) with special emphasis on descriptive sensory analysis (flavour profile method), J. Nutrit. Dietetics **1**, 219—260 (1964)

JELLINEK, G., Wissenschaftliche Grundlagen der sensorischen Geruchs- und Geschmacksanalyse, Gordian **68**, 9—14, 84—90, 132—135, 185—188, 228—230, 269—271 (1968)

JELLINEK, G., Sensorische Lebensmittelprüfung — Lehrbuch für die Praxis, Verlag Doris u. Peter Siegfried, D-3017 Pattensen/FRG 1981

JENNINGS, W. G. and M. R. SEVENANTS, Volatile esters of Bartlett pear, III., J. Food Sci. **29**, 158—163 (1964)

JENNINGS, W. and G. TAKEOKA, State-of-the-art fused silica capillary gas chromatography: Flavor problem applications, in P. SCHREIER, Analysis of volatiles, W. de Gruyter Berlin, New York 1984, p. 63—75

JIMENEZ, J. M. O., J. C. MARTIN and R. G. GONZALES-QUIJANO, Relación entre C. G. L.-perceptión sensorial de los aromas de aceites de oliva envasados, Grasas y Aceites **25**, 34—41 (1974)

JOUNELA-ERIKSSON, P., Whisky aroma evaluated by magnitude estimation, Lebensmittel-Wiss. u. -Technol. **15**, 303—308 (1982)

KADER, A. A., A. A. STEVENS, M. ALBRIGHT-HOLTON, L. L. MORRIS and M. ALGAZI, Effect of fruits ripeness when picked on flavour and composition in fresh market tomatoes, J. Amer. Soc. Hort. Sci. **102**, 724—731 (1977)

KAFKA, W. A., Specifity of odor-molecule interaction in single cells, in G. OHLOFF and A. F. THOMAS, Gustation and Olfaction, Academic Press London, New York 1971, p. 61—72

KAISER, R., Direkte und automatische Kopplung der Dünnschichtchromatographie an Gas-Chromatographen, Z. anal. Chemie **205**, 284—298 (1964)

KAMINSKI, E., S. STAWICKI, E. WASOWICZ, H. GIEBEL and R. PRZYBYLSKI, Geschmacks- und Geruchsverbindungen, die im Getreide und im Teig von Bakterien gebildet werden, Ber. 6. Tagung Internat. Probleme mod. Getreideverarbeit. u. Getreidechemie, Eigenverlag Inst. Getreideverarbeit. Bergholz-Rehbrücke 1975, p. 253—265

KAMINSKI, E., E. WASOWICZ, R. ZAWIRSKA and M. WOWER, The effect of drying and storage of dried carrots on sensory characteristics and volatile constituents, Nahrung **30** (1986), in press

KARE, M. R., Changes in taste with age — infancy to senescence, Food Technol. **29**, 8, 78 (1975)

KAREL, M., Recent research and development in the field of low-moisture and intermediate-moisture foods, Crit. Rev. Food Technol. **3**, 329—373 (1973)

KARLSSON-EKSTRÖM, G. and E. VON SYDOW, The aroma of black currants. VI. Correlations between instrumental and sensory data from unheated and heated black currants, Lebensmittel-Wiss. u. -Technol. **6**, 86—89 (1973)

KAWAMURA, Y. and A. ADACHI, Electrophysiological analysis of taste effectiveness of soda water and CO_2 gas, in T. HAYASHI, Olfaction and Taste, Pergamon Press London 1967, p. 431—437

KAYAHARA, K., N. AOYAGI, I. KAWASAKI and N. MIYACHI, Studies on flavor components of whiskey

IX. An analysis of whiskey malt and grain alcohol, J. Fermentat. Technol. [Japan] **44**, 633—639 (1966)

KENDALL, D. J. and A. J. NEILSON, Sensory and chromatographic analysis of mixtures formulated from pure odorants, J. Food Sci. **31**, 268—274 (1966)

KERKHOF, P. J. A. M. and H. A. C. THIJSSEN, The influence of processing on the structure of dehydrated liquid foods in relation to their aroma, Proc. Internat. Symposium on Aroma Research Zeist 1975, Centre Agric. Publish. and Doc. (Pudoc), Wageningen 1975, p. 167—192

KIERMEIER, F. and U. HAEVECKER, Zur sensorischen Qualitätsbeurteilung von Lebensmitteln. I. und II. Mitt., Z. Lebensmittel-Untersuch. u. -Forsch. **141**, 277—290 (1969)

KIERMEIER, F. and U. HAEVECKER, Sensorische Beurteilung von Lebensmitteln, Verlag J. F. Bergmann München 1972

KING, C. J. and H. A. MASSALDI, Aroma preservation during evaporation and drying of liquid foods: consideration of mechanism and effects of more than one phase being present. Work Doc. IV. Internat. Congr. Food Sci. and Technol. Madrid 1974, 5b, 3, p. 120

KOCH, J., Zur Technologie der Fruchtsaft-Herstellung, Flüssiges Obst **39**, 52—57 (1972)

KOCHAN, A., F. ÖRSI, P. MOLNAR and R. NEUMANN, Anwendung statistischer Tests zur Auswertung von Ergebnissen der sensorischen Analyse, Lebensmittel-Ind. **22**, 489—494 (1975)

KÖSTER, E. P. and O. WOUTERS, Entwicklung der sensorischen Methodik, Ernährungs-Umschau **17**, 349—354 (1970)

KOSIKOWSKI, F. V. and G. MOCQUOT, Advances in cheese technology, Food and Agric. Organisat., UN Agric. Studies no. 38 (1958) p. 141

KOVATS, E., Gaschromatographische Charakterisierung organischer Verbindungen. Teil 1: Retentionsindices aliphatischer Halogenide, Alkohole, Aldehyde und Ketone, Helv. chim. Acta **41**, 1915—1932 (1958)

KRAMER, C. Y., A method of choosing judges for a sensory experiment, Food Res. **20**, 492—496 (1955)

KRETOVIČ, V. L. and A. N. PONOMAREVA, Über die Beteiligung von Aminosäuren an der Reaktion der Melanoidinbildung beim Ausbacken von Brot [russ.], Biochimija **26**, 237—245 (1961)

KRÜGER, E. and L. NEUMANN, Beziehungen zwischen sensorischen und analytischen Merkmalen des Hopfens, Monatsschr. Brauerei **24**, 57—63 (1971)

KRÜGER, E. and L. NEUMANN, Die Zusammensetzung der nichtflüchtigen Aromastoff-Fraktion des Hopfens und die sich daraus ergebenden Beurteilungsmöglichkeiten, Monatsschr. Brauerei **28**, 20—27 (1975)

KURIHARA, K. and L. M. BEIDLER, Taste modifying protein from miracle fruit, Science **161**, 1241 1243 (1968)

LAFFORT, P., A linear relationship between olfactory effectiveness and identified molecular characteristics, in O. PFAFFMANN, Olfaction and Taste III, Rockefeller Univ. Press New York 1969, p. 150—157

LAND, D. G., Some factors influencing the perception of flavour-contributing substances in food, in D. G. LAND and H. E. NURSTEIN, Progress in flavour research, Appl. Sci. Publishers London 1979, p. 53—66

LAND, D. G., R. HARPER and N. M. GRIFFITHS, An evaluation of the odour qualities of some stimuli proposed as standards for odour research, Flavour Ind. **1**, 842—846 (1970)

LANGLER, J. E. and E. A. DAY, Development and flavour properties of methyl ketones in milk fat, J. Dairy Sci **47**, 1291—1296 (1964)

LANGSRUD, T. and G. W. REINBOLD, Flavour development and microbiology of Swiss cheese — A review, J. Milk and Food Technol. **36**, 487—490, 531—542, 593—609 (1973); **37**, 26—41 (1974)

LARMOND, E., Physical requirements for sensory testing, Food Technol. **27**, 11, 28—32 (1973)

LATRASSE, A., J. R. DEGORCE-DUMAS and J. Y. LEVEAU, Production d' aromês par les microorganismes, Sci. Aliments **5**, 1—26, 1985.

LEAHY, M. M. and G. A. REINECCIUS, Comparison of methods for the isolation of volatile compounds from aqueous model systems, in P. SCHREIER, Analysis of volatiles, Walter de Gruyter Berlin/New York 1984, p. 19—47

LEMBERGER, F., Psychophysische Untersuchungen über den Geschmack von Zucker und Saccharin, Pflügers Arch. ges. Physiol. **123**, 293—311 (1908)

LILJESTRAND, G. and Y. ZOTTERMAN, The alcaline taste, Acta physiol. scand **35**, 380—389 (1956)

LILLARD, D. A., M. MONTGOMERY and E. A. DAY, Flavor threshold values of certain carbonyl compounds in milk, J. Dairy Sci. **45**, 660 (1962)

LINDSAY, R. C., Cultured dairy products, in H. W. SCHULTZ, E. A. DAY and L. M. LIBBEY, Symposium on Foods: The Chemistry and Physiology of Flavors, The Avi Publ. Co. Westport/Conn. 1967, p. 315—330

LINKO, Y.-Y. and J. A. JOHNSON, Changes in amino acids and formation of carbonyl compounds during baking, J. agric. and Food Chem. **11**, 150—152 (1963)

LOVELOCK, J. E. and S. R. LIPSKY, Electron affinity spectroscopy — a new method for the identification of functional groups in chemical compounds separated by gas chromatography, J. Amer. chem. Soc. **82**, 431—433 (1960)

LUNDGREN, B., B. JONSSON, R. M. PANGBORN, A. M. SONTAG, N. BARYLKO-PIKIELNA, E. PIETRZAK, R. DOS SANTOS GARRUTI, M. A. CHAIB MORAES and M. YOSHIDA, Taste discrimination by hedonic response to sucrose in coffee beverages. An interlaboratory study, Chem. Senses and Flavour **3**, 249—265 (1978)

MAARSE, H., Introduction, in H. MAARSE and R. BELZ, Isolation, separation and identification of volatile compounds in aroma research, Akademie-Verlag Berlin 1981, p. 1

MAARSE, H., Methods of separating and identifying flavour compounds, in L. NYKÄNEN and P. LEHTONEN, Flavour research of alcoholic beverages, Foundat. Biotechn. and Ind. Res. Helsinki **3**, 71—97 (1984)

MAARSE, H. and M. C. TEN NOEVER DE BRAUW, The analysis of volatile components of Jamaica Rum, J. Food Sci, **31**, 951—955 (1966)

MAGA, J. A., Flavor potentiators, Crit. Rev. Food Sci. and Nutrit. **18**, 231—312 (1983)

MAGA, J. A. and C. E. SIZER, Pyrazines in foods, Crit. Rev. in Food Technol. **4**, 39—115 (1973)

MAIER, H. G., Flüchtige Aromastoffe in Lebensmitteln, Angew. Chemie **82**, 965—974 (1970)

MAIER, H. G., Bindung flüchtiger Aromastoffe an Lebensmittel, Lebensmittel-Wiss. u. -Technol. **5**, 1—6 (1972)

MAIER, H. G., Zur Bindung flüchtiger Aromastoffe an Proteine, Dt. Lebensmittel-Rdsch. **70**, 349—351 (1974)

MAIER, H. G., Binding of volatile aroma substances to nutrients and foodstuffs, Proc. Internat. Symposium on Aroma Research Zeist/Netherlands 1975, Centre Agric. Publish. and Doc. (Pudoc) Wageningen 1975, p. 143—157

MARTIN, S. L., Selection and training of sensory judges, Food Technol. **27**, 11, 22—26 (1973)

MAULE, D. R., Rapid gaschromatographic examination of beer flavour, J. Inst. Brew. **73**, 351—361 (1967)

MAYER, R. and F. WITTIG, Über einen Zusammenhang zwischen bitterem Geschmack und chemischer Konstitution von Thiocarbonylverbindungen, Z. Chemie **12**, 91—100 (1972)

MCCARTHY, A. J., J. K. PALMER, C. P. SHAW and E. E. ANDERSON, Correlation of gas chromatographic data with flavor profiles of fresh banana fruit, J. Food Sci. **28**, 379—384 (1963)

MCFADDEN, W. H., R. TERANISHI, D. R. BLACK and J. C. DAY, Use of capillary gas chromatography with a time-of-flight mass spectrometer, J. Food Sci. **28**, 316—319 (1963)

MCKERNAN, W. M. Microencapsulation in the flavour industry, I., Flavour-Ind. **3**, 596—600 (1972)

MCLEAD, W. D. and N. M. BUIGUES, Sesquiterpenes. I. Nootkatone, a new grapefruit flavor constituent, J. Food Sci. **29**, 565—568 (1964)

MCWILLIAM, I. G. and R. A. DEWAR, Flame ionization detector for gas chromatography, Nature [London] **181**, 760 (1958)

MECREDY, J. M., J. C. SONNEMANN and S. J. LEHMANN, Sensory profiling of beer by a modified QDA method, Food Technol. **28**, 11, 36—41 (1974)

MEILGAARD, M. C., Aroma volatiles in beer: Purification, flavour, threshold and interaction, in F. DRAWERT, Geruch- und Geschmackstoffe, Internat. Symposium der Fa. Haarmann & Reimer Bad Pyrmont 1974, Verlag Hans Carl Nürnberg 1975, p. 211—254

MILLEVILLE, H. P. and R. K. ESKEW, (Western) Canner and Packer **1946**, 10, 51

MILUTINOVIC, L., R. E. BARGMANN, K. CHANG, M. CHASTAIN and J. J. POWERS, Comparison of flavor and volatiles of tomato products and of peanuts, J. Food Sci, **35**, 224—228 (1970)

MOHLER, H., Enzyme bei der Bildung von Aromastoffen, in J. SCHORMÜLLER, Handbuch der Lebensmittelchemie, Bd. I, Springer-Verlag Berlin, Heidelberg, New York 1965, p. 912

MOHR, W., E. LANDSCHREIBER and T. SEVERIN, Zur Spezifität des Kakaoaromas, Fette, Seifen, Anstrichmittel **78**, 88—95 (1976)

MONCRIEFF, R. W., Gustation, in R. W. MONCRIEFF, The Chemical Senses, Verlag Leonard Hill London 1967, p. 269—270

MOSANDL, A. and G. HEUSINGER, Stereoisomers of fruit flavour substances — some aspects of synthesis and analysis, in P. SCHREIER, Analysis of volatiles, Walter de Gruyter Berlin/New York 1984, p. 343—356

MOSEL, J. N. and KANTOROWITZ, The effect of monosodium glutamate on acuity to the primary tastes, Amer. J. Psychol. **65**, 573—579 (1952)

MOSHONAS, M. G. and P. E. SHAW, Direct gas chromatographic analysis of aqueous citrus and other fruit essences, J. agric. and Food Chem. **32**, 526—530 (1984)

MOSKOWITZ, H. R., Combination rules for judgement of odor quality difference, J. agric. and Food Chem. **22**, 740—743 (1974)

MOSKOWITZ, H. R., Sensory evaluation by magnitude estimation, Food Technol. **28**, 11, 16 (1974)

MOSKOWITZ, H. R., Applications of sensory measurement to food evaluations. II. Methods of ratio scaling, Lebensmittel-Wiss. u. -Technol. **8**, 249 (1975)

MOSKOWITZ, H. R., Magnitude estimation: Notes on how, what, where and why to use it, J. Food Qual. **1**, 195 (1978)

MULDERS, E. J., The odour of white bread. IV. Quantitative determination of constituents in the vapour and their odour values, Z. Lebensmittel-Untersuch. u. -Forsch. **151**, 310—317 (1973)

MOSKOWITZ, H. R., V. KUMARIAH, K. N. SHARMA, H. L. JACOBS and S. D. SHARMA, Cross-cultural differences in simple taste preference, Science **19**, 1217—1218 (1975)

MULDERS, E. J. and C. WEURMAN, Investigation of the aroma of white bread, Ber. 5. Welt-Getreide- u. Brotkongreß Dresden 1970, VEB Fachbuchverlag Leipzig 1970, 5, p. 215—219

MURDOCK, D. I., Diacetyl test as a quality control tool in processing frozen concentrated orange juice, Food Technol. **22**, 90—94 (1968)

NAGY, S. and H. L. DINSMORE, Relationship of furfural to temperature abuse and flavour change in commercially canned single-strength orange juice, J. Food Sci. **39**, 1116—1119 (1974)

NAWAR, W. W., Some considerations in interpretation of direct headspace gas chromatographic analyses of food volatiles, Food Technol. **20**, 213—215 (1966)

NELSON, J. H., Production of Blue cheese flavor via submerged fermentation by Penicillium roqueforti, J. agric. and Food Chem. **18**, 567—569 (1970)
NELSON, P. E. and J. E. HOFF, Food volatiles: Gas chromatographic determination of partition coefficients in water-lipid systems, J. Food Sci. **33**, 479—482 (1968)
NEUMANN, R., Zur Vertrauenswürdigkeit sensorischer Analysenergebnisse, Lebensmittel-Ind. **26**, 109—112 (1979); **28**, 7—11 (1981)
NEUMANN, R., S. ARNOLD and P. MOLNAR, Zur Standardisierung der sensorischen Erzeugnisprüfung, Lebensmittel-Ind. **21**, 352—360 (1974)
NEY, K. H., Aromaintensivierende Wirkung von L-Glutamat und ähnlichen Verbindungen, Z. Lebensmittel-Untersuch. u. -Forsch. **146**, 141—143 (1971)
NEY, J. H., Aminosäure-Zusammensetzung von Proteinen und die Bitterkeit ihrer Peptide, Z. Lebensmittel-Untersuch. u. -Forsch. **149**, 321—323 (1972)
NITZ, S. and E. JÜLICH, Concentration and GC-MS analysis of trace volatiles by sorption-desorption techniques, in P. SCHREIER. Analysis of volatiles, W. de Gruyter Berlin/New York 1984, p. 151—170
NOBLE, A. C., A chemical approach to aroma, Cereal foods world **21**, 59—61 (1976)
TEN NOEVER DE BRAUW, M. C. and C. BRUNNEE, Ergebnisse mit einer optimierten Kopplung von Gaschromatograph und Massenspektrometer, Z. anal. Chemie **229**, 321—335 (1967)
TEN NOEVER DE BRAUW, M. C. and C. VAN INGEN, Mass spectrometry, in H. MAARSE and R. BELZ, Isolation, separation and identification of volatile compounds in aroma research, Handbuch der Aromaforschung, Akademie-Verlag Berlin 1981, p. 155
NONAKA, M., A new procedure for isolation poultry aroma essence, Food Technol. **25**, 1122—1124 (1971)
NOVOTNY, M., M. L. MCCONNELL and M. L. LEE, Some aspects of high resolution gas chromatographic analysis of complex volatile samples, J. agric. and Food Chem. **22**, 765—770 (1974)
NUNEZ, A. J., J. M. H. BEMELMANS and H. MAARSE, Isolation methods for the volatile components of grapefruit juice. Distillation and solvent extraction methods, Chromatographia **18**, 153—158 (1984)
NURSTEN, H. E., Why flavour research? How far have we come since 1975 and where now?, in D. G. LAND and H. E. NURSTEN, Progress in Flavour Research, Appl. Sci. Publishers London 1979, P. 337—355
NURSTEN, H. E., Chemistry of flavours. Past, present and future, Flavours **1975**, 75—82
NURSTEN, H. E. and M. L. WOOLFE, An examination of the volatile compounds present in cooked Bramley's seedling apples and the changes they undergo in processing, J. Sci. Food and Agric. **23**, 803—822 (1972)
NYKÄNEN, L. and H. SUOMALAINEN, Formation of aroma compounds by yeast. Dicarbonyl compounds, in L. NYKÄNEN and H. SUOMALAINEN, Aroma of beer, wine and distilled alcoholic beverages, Handbuch der Aromaforschung, Akademie-Verlag Berlin 1983, p. 5—7, 68—71
ÖRSI, F., Mathematisch-statistische Auswertung der Ergebnisse der sensorischen Punktbewertung, Lebensmittel-Ind. **20**, 103—107 (1973)
OLNEY, J. W., Brain lesions, obesity, and other disturbances in mice treated with monosodium glutamate, Science **164**, 719—721 (1969)
OLNEY, J. W. and L. G. SHARPE, Monosodium glutamate: specific brain lesions questioned, Science **167**, 1017 (1970)
OLSEN, R. D., 15 ppm Maltol replaces 5—15 percent sugar in beverages, Food Process. **25**, 1, 95 (1964)
OPDYKE, D. L. J., Monographs on fragrance raw materials, Food and Cosmetics Toxicol. **11**, 95—115, 447—495, 855—876, 1011—1081 (1973); **12**, 385—405, 517—537 (1974); **13**, 91—112, 449—457 (1975)
OSER, B. L., K. MORGAREIDGE and S. CARSON, Monosodium glutamate studies in 4 species of neonatal and infant animals, Food and Cosmetics Toxicol. **13**, 7—14 (1975)
OSMAN-ISMAIL, F. and J. SOLMS, Formation of inclusion compounds of starches with flavour substances, Lebensmittel-Wiss. u. -Technol. **6**, 147—150 (1973)
OUGH, C. S. and H. STONE, An olfactometer for rapid and critical odor measurement, J. Food Sci. **26**, 452—456 (1961)
OUPADISSAKOON, C. and C. T. YOUNG, Modeling of roasted peanut flavor for some Virginia-type peanuts from amino acid and sugar contents, J. Food Sci. **49**, 52—58 (1984)
PAILLARD, N., S. PITOULIS and A. MATTEL, Techniques de préparation et analyse de l'arôme de quelques fruits, Lebensmittel-Wiss. u. -Technol. **3**, 107—114 (1970)
PANGBORN, R. M. and S. C. OUGH, Techniques for sensory evaluation of food odors, Advances Food Res., Acad. Press New York, London 1965, XIV, p. 1—32
PANGBORN, R.-M. and A. S. SZCZESNIAK, Effects of hydrocolloids and viscosity on flavour- and odour-intensities of aromatic flavour compounds, J. Texture Studies **4**, 467—482 (1974)
PATTEE, H, and J. SINGLETON, Comparison of geographical and varietal effects on the peanut volatile profile by peak-ratio analysis, J. agric. and Food Chem. **20**, 1119—1124 (1972)
PATTON, S., The mechanism of sunlight flavor formation in milk with special reference to methionine and riboflavine, J. Dairy Sci. **37**, 446—449 (1954)
PATTON, S. and D. V. JOSEPHSON, A method for determining significance of volatile flavor compounds in foods, Food Res. **22**, 316—318 (1957)
PAULING, L., Chem. Engng. News **24**, 1064—1065 (1946)
PEER, H. C. and B. HOOGSTAD, Method for encapsulation of polar compounds in food. Proc. Internat. Symposium on Aroma Research Zeist/Netherlands 1975, Centre for Agric. Publ. and Doc. (Pudoc) Wageningen 1975, p. 159—166
PERSSON, T. and E. VON SYDOW, Aroma of canned beef: gas chromatographic and mass spectrometric

analysis of the volatiles, J. Food Sci. **38**, 377—385 (1973)

PERSSON, T., E. VON SYDOW and C. ÅKESON, Aroma of canned beef: sensory properties, J. Food Sci. **38**, 386—392 (1973)

PESARO, M., G. BOZZATO and P. SCHUDEL, The total synthesis of racemic nootkatone, Chem. Commun. **1968**, 19, 1152—1154

PETRITSCHEK, A., F. LYNEN and H. D. BELITZ, Untersuchungen über Bitterpeptide. II. Auftreten von Bittergeschmack in enzymatischen Hydrolysaten verschiedener Proteine, Lebensmittel-Wiss. u. -Technol. **5**, 77—81 (1972)

PETUELY, F., Lebensmittelrechtliche Regelung der Aromatisierung in Österreich, in W. AUERSWALD und B. M. BRANDSTETTER, Natürliche, natur-identische und synthetische Geschmacksstoffe und deren lebensmittelrechtliche Aspekte, Wilhelm Maudrich Verlag Wien 1973, p. 23—26, 33—34

PINO, J., R. TORRICELLA and F. ÖRSI, Correlation between sensory and gas chromatographic measurements on grapefruit juice volatiles, Nahrung **30**, 783—790 (1986)

PINTAURO, N., Flavor Technology 1971, Noyes Data Corporation Park Ridge/N. J. 1971, p. 1

POKORNÝ, J., H. BULANTOVA and J. DAVÍDEK, Sensory profiles of browning products from reaction of 2-furaldehyde with L-lysine, Nahrung **25**, 531—541 (1981)

POKORNÝ, J., L. NGUYEN-THIEN, B. A. EL ZEANY and G. JANÍČEK, Bildung von Fischaroma durch Reaktion autoxydierter Lipide mit Aminosäuren und Proteinen, Nahrung **20**, 273—279 (1976)

PONOMAREVA, A. N., V. L. KRETOVIČ, I. I. KAREVA and T. JAKUBCZYK, Dynamik des Gehaltes an freien Aminosäuren im Verlauf des Produktionsprozesses von Weizenbrot [russ.], Biochimija **29**, 283—287 (1964)

POWERS, J. J., R. J. COX, M. C. ELLAND and P. ACKROYD-KELLY, Exposure of peanuts and peanut volatiles to light: influence on flavour and gas chromatographic profile, Flavour-Ind. **2**, 87—92 (1971)

POWERS, J. J. and E. S. KEITH, Stepwise discriminant analysis of gas chromatographic data as an aid in classifying the flavor quality of foods, J. Food Sci. **33**, 207—213 (1968)

POWERS, J. J. and M. C. QUINLAN, Refining of methods for subjective-objective evaluation of flavor, J. agric. and Food Chem. **22**, 744—749 (1974)

PYKE, M., A taste for meat, Internat. Flavours and Food Additives **6**, 275—277 (1975)

REINECCIUS, G. A., P. G. KEENEY and W. WEISSBERGER, Factors affecting the concentration of pyrazines in cocoa beans, J. agric. and Food Chem. **20**, 202—206 (1972)

REINER, L. and A. PIENDL, Zur Gruppierung von Biereigenschaften mittels der Faktoranalyse, Brauwiss. **27**, 1—10 (1973)

RIJKENS, F. and H. BOELENS, The future of flavour research, Proc. Internat. Symposium on Aroma Research Zeist/Netherlands 1975, Centre Agric. Publish. and Doc. (Pudoc) Wageningen 1975, p. 203—220

RIKLIN, R., Anwendung von Aromakonzentraten in verarbeiteten Lebensmitteln, in J. SOLMS und H. NEUKOM, Aroma- und Geschmacksstoffe in Lebensmitteln, Forster-Verlag Zürich 1967, p. 277—288

RÖDEL, W., Möglichkeiten einer enzymatischen Aromaverbesserung bei Obst und Gemüse, Nahrung **15**, 431—438 (1971)

RÖDEL, W., D. ZÖLL and G. WÖLM, Beitrag zur Anreicherung flüchtiger Stoffe aus verdünnten wäßrigen Lösungen für die gaschromatographische Analyse, Nahrung **15**, 425—429 (1971)

ROTHE, M., Über flüchtige Aromastoffe des Roggenbrotes, Ernährungsforschung **5**, 131—142 (1960)

ROTHE, M., Zur Frage der Geruchsentwicklung in fadenziehendem Brot, Ernährungsforschung **8**, 432—439 (1963)

ROTHE, M., Aktuelle Probleme der modernen Aromaforschung, Nahrung **18**, 115—123 (1974)

ROTHE, M., Chemie und Biochemie der Aromabildung in Brot, in M. ROTHE, Aroma von Brot, Handbuch der Aromaforschung, Akademie-Verlag Berlin 1974, p. 17—44

ROTHE, M., W. ENGST and V. ERHARDT, Studies on characterization of Blue cheese flavour, Nahrung **26**, 591—602 (1982)

ROTHE, M., W. ENGST and M. SPECHT, Beispiele für die Lösung von Aromaproblemen mittels sensorischer Profilanalysen, Chemie, Mikrobiol., Technol. Lebensmittel **7**, 13—17 (1981)

ROTHE, M., G. REHFELD and L. TUNGER, Ernährungswissenschaftliche Betrachtungen zum Backprozeß in der Dampfbackkammer, Ernährungsforschung **12**, 215—229 (1967)

ROTHE, M. and H. RUTTLOFF, Aroma retention in modern bread production, Nahrung **27**, 505—512 (1983)

ROTHE, M. and M. SPECHT, Bemerkungen zum Molekulargewicht von Aromastoffen, Nahrung **20**, 281—286 (1976)

ROTHE, M. and J. STÖCKEL, Der Fettsäuregrad als Maßzahl zur Beurteilung von Getreide, Getreidemühle **5**, 156—157, 182—183 (1961)

ROTHE, M., J. STÖCKEL and R. HECKEL, Gesundheitszustand und Lipaseaktivität von Roggen, Ernährungsforschung **12**, 265—271 (1967)

ROTHE, M. and B. THOMAS, Aromastoffe des Brotes. Versuch einer Auswertung chemischer Geschmacksanalysen mit Hilfe des Schwellenwertes, Z. Lebensmittel-Untersuch. u. -Forsch. **119**, 302—310 (1963)

ROTHE, M. and L. TUNGER, Getreideerzeugnisse in der heutigen Ernährung, Ber. 6. Tagung Internat. Probleme mod. Getreideverarb. u. Getreidechemie, Eigenverlag Inst. Getreideverarb. Bergholz-Rehbrücke 1975, p. 1—14

ROTHE, M. and G. WÖLM, Analytische Bewertung der Qualität von Weinbrand, Ernährungsforschung **15**, 309—318 (1970)

ROTHE, M., G. WÖLM, L. TUNGER and H.-J. SIEBERT, Schwellenkonzentrationen von Aromastoffen und ihre Nutzung zur Auswertung von Aromaanalysen, Nahrung **16**, 483—495 (1972)

ROTHE, M., G. WÖLM and I. VOIGT, Fettveränderungen in Getreideprodukten und ihr analytischer Nachweis. 1. Mitt. Ranzigkeit, Nahrung **11**, 149—160 (1967)

RYHAGE, R., Use of mass spectrometer as a detector and analyzer for effluents emerging from high temperature gas liquid chromatography columns, Analyt. Chem. **36**, 759—764 (1964)

SALO, P., Determining the odor threshold for some compounds in alcoholic beverages, J. Food Sci. **35**, 95—99 (1970)

SALO, P., Interactions among volatile aroma compounds in a whisky imitation, Lebensmittel-Wiss. u. Technol. **6**, 52—58 (1973)

SALO, P., Use of odour threshold in sensorial testing and comparisons with instrumental analysis, Proc. Internat. Symposium on Aroma Research Zeist/ Netherlands 1975, Centre Agric. Publish. and Doc. (Pudoc) Wageningen 1975, p. 121—130

SALO, P., L. NYKÄNEN and H. SUOMALAINEN, Odor thresholds and relative intensities of volatile aroma components in an artificial beverage imitating whisky, J. Food Sci. **37**, 394—398 (1972)

SAPERS, G. M., J. F. SULLIVAN and F. B. TALLEY, Flavor quality in explosion puffed dehydrated potato. 1. A gas chromatographic method for the determination of aldehydes associated with flavor quality. 2. Flavor contribution of 2-methylpropanal, 2-methylbutanal and 3-methylbutanal, J. Food Sci. **35**, 728—733 (1970)

SATO, M., N. AKAIKE and S. YAMASHITA, Electrophysiological studies on the effects of ribonucleotides on taste receptors of rats. Effects on the addition of ribonucleotides to monosodium glutamate and of temperature, Symp. Amino Acid Nucleic Acid **11**, 53—61 (1965)

SAUNDERS, D. L. and L. R. SNYDER, Drum TLC. A new technique for rapid, high efficiency separations, J. chromatogr. Sci. **8**, 706—716 (1970)

SCHAEFER, J., Isolation and concentration of volatiles from foods, Isolation and concentration from the vapour phase, in H. MAARSE and R. BELZ, Isolation, separation and identification of volatile compounds in aroma research, Handbuch der Aromaforschung, Akademie-Verlag Berlin 1981, p. 37—50

SCHAEFER, J. and C. WEURMAN, Einige Aspekte der gaschromatographischen Beurteilung der Qualität von Lebensmitteln, Ernährungsforschung **15**, 301—307 (1970)

SCHEIDE, J., Aroma und Medium. Wechselwirkungen und Zusammenhänge, H & R contact, Schriftenreihe der Fa. Haarmann & Reimer Holzminden/ BRD 1975, 12, p. 7—12

SCHLEGEL, W., Die Anwendung von Aromastoffen in Lebensmitteln, Gordian **68**, no. 1611, 15—19 (1968)

SCHLEGEL, W., Nutrition and flavour-legislation, Nahrung **20**, 253—258 (1976)

SCHMIDT, H. W. H., Anwendung der Dünnschicht-Chromatographie bei der Erforschung und Analyse von Aroma- und und Geschmacksstoffen, in J. SOLMS und H. NEUKOM, Aroma- und Geschmacksstoffe in Lebensmitteln, Forster-Verlag Zürich 1967, p. 35—46

SCHNEIDER, D., Insect olfaction: deciphering system for chemical messages, Science **163**, 1031—1037 (1969)

SCHNEIDER, D., Molekulare Grundlagen der chemischen Sinne bei Insekten, Naturwiss. **58**, 194—200 (1971)

SCHNEIDER, D., Specialized odor receptors of insects, in G. OHLOFF and A. F. THOMAS, Gustation and Olfaction, Academic Press London, New York 1971, p. 45—60

SCHNEIDER, D., V. LACHER and K.-E. KAISSLING, Die Reaktionsweise und das Reaktionsspektrum von Riechzellen bei *Antheraea pernyi*, Z. vergl. Physiol. **48**, 632—662 (1964)

SCHÖNBERG, A. and R. MOUBACHER, The Strecker-degradation of α-amino acids, Chem. Rev. **50**, 261—277 (1952)

SCHÖNBERG, A. R. MOUBACHER and A. MOSTOFA, Degradation of α-amino acids to aldehydes and ketones by interaction with carbonyl compounds, J. Chem. Soc. **1948**, 176—182

SCHOMBURG, G., H. HUSMANN, L. PODMANICZYK, F. WEEKE and A. RAPP, Coupled gas chromatographic methods for separation, identification, and quantitative analysis of complex mixtures: MDGC, GC-MS, GC-IR, LC-GC, in P. SCHREIER, Analysis of volatiles, W. de Gruyter Berlin/New York 1984, p. 121—150

SCHREIER, P., Analysis of volatiles, W. de Gruyter Berlin/New York 1984

SCHRÖDTER, R., Die gas-chromatographische Analyse von Aromastoffen in Korrelation zu sensorischen Daten am Beispiel des Apfelsaftes, Diss. Techn. Univ. Dresden 1976

SCHUTTE, L., Precursors of sulfur-containing flavor compounds, Crit. Rev. Food Technol. **4**, 457—505 (1974)

SCHUY, K. D. and D. H. HUNNEMAN, Kombinierte Gas-Chromatographie/Massenspektrometrie — Grundlagen des Verfahrens und Bedeutung der Kopplung in der Chemie, Chemiker-Z. **95**, 633—643 (1971)

SCHWARTZ, D. P. and O. W. PARKS, Methods for the isolation and characterization of constituents of natural products, Microchem. J. **1963**, 7, 403

SCHWIMMER, S., D. W. VENSTROM and D. G. GUADAGNI, Relation between pyruvate content and odor strenght of reconstituted onion powder, Food Technol. **18**, 121—124 (1964)

SCHWIMMER, S. and W. J. WESTON, Enzymatic development of pyruvic acid in onion as a measure of pungency, J. agric. and Food Chem. **9**, 301—304 (1961)

SEIBEL, W. and H. STEPHAN, Zur Entwicklung des Brotsortiments in den 70er Jahren, Brot u. Gebäck **24**, 61—66 (1970)

SEIFERT, R. M., R. G. BUTTERY, D. G. GUADAGNI, D. R. BLACK and J. G. HARRIS, Synthesis of some 2-methoxy-3-alkylpyrazines with strong bell pepper-

like odors, J. agric. and Food Chem. **18**, 246—249 (1970)
SEIFERT, R. M., R. G. BUTTERY, D. G. GUADAGNI, D. R. BLACK and J. G. HARRIS, Synthesis and odor properties of some additional compounds related to 2-isobutyl-3-methoxypyrazine, J. agric. and Food Chem. **20**, 135—137 (1972)
SEITZ. E. W., Industrial application of microbial lipases: a review, J. Amer. Oil Chemist's Soc. **51**, 12—16 (1974)
SHALLENBERGER, R. S. and T. A. ACREE, Molecular theory of sweet taste, Nature **216**, 480—482 (1967)
SHAPIRO, J., Freezing-out, a safe technique for concentration of dilute solutions, Science **133**, 2063—2064 (1961)
SHAPIRO, J., Concentration of volatile substances in aqueous solution and production of water free of organics by freezing-out, Analyt. Chem. **39**, 280 (1967)
SHIBAMOTO, T., Application of high resolution capillary columns on flavor and fragrance analysis, in P. SCHREIER, Analysis of volatiles, W. de Gruyter Berlin/New York 1984, p. 233—250
SHIPE, W. F., R. BASSETTE, D. D. DEANE, W. L. DUNKLEY, E. G. HAMMOND, W. J. HARPER, D. H. KLEYN, M. E. MORGAN, J. H. NELSON and R. A. SCANLAN, Off-flavors of milk: nomenclature, standards, and bibliography, J. Dairy Sci. **61**, 855—869 (1978)
SIEK, T. J., I. A. ALBIN, L. A. SATHER and R. C. LINDSAY, Taste thresholds of butter volatiles in deodorized butteroil medium, J. Food Sci. **34**, 265—267 (1969)
VON SKRAMLIK, E., Über die Giftigkeit und Verträglichkeit von ätherischen Ölen, Pharmazie **14**, 435—445 (1959)
SMITH, K. H., Savoury seasonings for meat and snack food, Food Process. Ind. **43**, 516, 40—47 (1974)
SOLMS, J., Geschmacksaktive Verbindungen, vor allem in Fleisch und Gemüse, in J. SOLMS und H. NEUKOM, Aroma- und Geschmacksstoffe in Lebensmitteln, Forster-Verlag Zürich 1967, p. 199—221
SOLMS, J., Die Bedeutung von Lebensmittelaromen, in W. AUERSWALD und B. M. BRANDSTETTER, Natürliche, natur-identische und synthetische Geschmacksstoffe und deren lebensmittelrechtliche Aspekte, Verlag Wilhelm Maudrich Wien 1973, p. 5—10
SOLMS, J., Sensorische Aspekte bei synthetischen Lebensmitteln, Internat. Z. Vitamin- u. Ernähr.-Forsch., Beiheft Nr. 14, Qualitätskriterien der Nahrung **1974**, p. 191—199
STAHL, E., Thin-layer chromatography, Pharmazie **11**, 633—637 (1956)
STEINER, W., Probleme der Geruchsforschung heute, Chemiker-Z. **97**, 3—8 (1973)
STEINSHOLT, K. and A. SVENSEN, Catty flavour in cheese resulting from the addition of trace amounts of mesityl oxide to cheese milk, Milchwiss. **34**, 598—599 (1979)
STENHAGEN, E., S. ABRAHAMSSON and F. W MCLAFFERTY, Atlas of mass spectral data, Interscience Publishers New York 1969
STEVENS, S. S., The direct estimation of sensory magnitudes: loudness, Amer. K. Psychol. **69**, 1—25 (1956)
STÖCKLIN, W., E. WEISS and T. REICHSTEIN, Gymnemasäure, das antisaccharine Prinzip von *Gymnema sylvestre R. Br.* Isolierungen und Identifizierungen, Helv. chim. Acta **50**, 474—490 (1967)
STOLL, A. and E. SEEBECK, Chemical investigations on alliin, the specific principle of garlic, Advances Enzymol. **11**, 377—400 (1951)
STONE, H., J. SIDEL, S. OLIVER, A. WOOLSEY and R. C. SINGLETON, Sensory evaluation by Quantitative Descriptive Analysis, Food Technol. **28**, 11, 24—34 (1974)
VAN STRATEN, S. and H. MAARSE, Volatile compounds in food. Qualitative data, Div. Nutrit. and Food Res. TNO, Zeist/Netherlands, 5th edit. and supplements 1983, 1984
VAN STRATEN, S. and F. J. G. DE VRIJER, Lists of volatile compounds in food, Central Institute of Food and Nutrition, Zeist/Netherlands 1973, Report Nr. R 4030
STREULI, H., Röstaroma, in J. SOLMS und H. NEUKOM, Aroma- und Geschmacksstoffe in Lebensmitteln, Forster-Verlag Zürich 1967, p. 123
STUIVER, M., Biophysics of the sense of smell, Diss. Univ. Groningen/Netherlands 1958
SAGUWARA, S., U. KOBASHI and H. SAKURAI, Studies on the chemical evaluation of tobacco quality. Part. I. Possibility for the evaluation of tobacco quality from chemical composition in particulate phase of the cigarette smoke, Agric. and biol. Chem. **35**, 596—603 (1971)
SUGISAWA, H., Prerequisites for sample preparation in flavour analysis, in P. SCHREIER, Analysis of volatiles, W. de Gruyter Berlin/New York 1984, p. 3—18
SUGISAWA, H., C. CHOKOU and N. KENSUKE, Simultaneous distillation-adsorption and its application, in P. SCHREIER, Analysis of volatiles. W. de Gruyter Berlin/New York 1984, p. 357—369
SUOMALAINEN, H. and L. NYKÄNEN, Methods applied in studies on the aroma composition of alcoholic beverages, Wallerstein Lab. Comm. **31**, no. 104, 5—15 (1968)
SUOMALAINEN, H. and L. NYKÄNEN, Composition of whisky flavour, Process Biochem. **1970**, 1—6
SUOMALAINEN, H. and L. NYKÄNEN, Formation of aroma compounds in alcoholic beverages, Wallerstein Lab. Commun. **35**, 185—202 (1972)
VON SYDOW, E., Flavor — a chemical or psychophysical concept?, Food Technol. **25**, 1, 40—44 (1971)
VON SYDOW, E., Contribution of individual chemical compounds to the aroma of foods, in F. DRAWERT, Geruch- und Geschmacksstoffe, Internat. Symposium der Fa. Haarmann & Reimer Bad Pyrmont 1974, Verlag Hans Carl Nürnberg 1975, p. 187—200
VON SYDOW, E., J. ANDERSSON, K. ANJOU, G. KARLSSON, D. G. LAND and N. M. GRIFFITHS, The aroma of bilberries (Vaccinium myrtill s L.), II. Evaluation of the press juice by sensory methods and by

gas chromatography and mass spectrometry, Lebensmittel-Wiss. u. -Technol. **3**, 11—17 (1970)
VON SYDOW, E. and G. KARLSSON, The aroma of black currants. IV. The influence of heat measured by instrumental methods, Lebensmittel-Wiss. u. Technol. **4**, 54—58 (1971)
TÄUFEL, K. and R. VOGEL, Filtrierpapier-Test zum Nachweis peroxydischer Verbindungen, Fette u. Seifen **57**, 393—399 (1955)
TELEGDY-KOVATS, L., F. ÖRSI and M. ÖRSI-JEZSIK, Theoretische Fragen zur Auswertung von sensorischen Punktbewertungsverfahren von Lebensmitteln, Nahrung **14**, 531—545 (1970)
TERANISHI, R., I. HORNSTEIN, P. ISSENBERG and E. L. WICK, Mass spectrometry: Instrument requirements and limitations, in Flavor Research — Principles and Techniques, Verlag Marcel Dekker New York 1971, p. 133—179
THEALL, C. L. and W. L. CHENOWETH, Initial food choice and sucrose preference in offspring of rats fed starch-or sucrose, Nutrit. Reports Internat. **28**, 1301—1312 (1983)
THEIMER, E. T. and J. T. DAVIES, Olfaction, musk odor, and molecular properties, J. agric. and Food Chem. **15**, 6—14 (1967)
TILGNER, D. J., Anchored sensory evaluation tests — a status report, Food Technol. **16**, 3, 47—50 (1962)
TILGNER, D. J., Dilution tests for odor and flavor analysis, Food Technol. **16**, 2, 26—29 (1962)
TILGNER, D. J., A retrospective view of sensory analysis and some considerations for the future, Adv. Food Res. **19**, 215—277 (1971)
TILGNER, D. J., Fortschritte in der sensorischen Gütebewertung, Gordian **74**, 238—242 (1974)
TOLLENAAR, F. D., Qualitative factors in food production, Vortrag 8. TNO-Konferenz 1975, Central Institute of Food and Nutrition Zeist/Netherlands 1975
TRACEY, M. V., Foods of the future, Food Technol. Australia **21**, 8—11 (1969)
TRANT, A. S. and R. M. PANGBORN, Discrimination, intensity, and hedonic response to color, aroma, viscosity, and sweetness of beverages, Lebensmittel-Wiss. u. -technol. **16**, 147—152 (1983)
TREIBS, W., Die ätherischen Öle, I. Geschichtliche Einleitung, in E. GILDEMEISTER and F. HOFFMANN, Die ätherischen Öle, Akademie-Verlag Berlin 1956, 1, p. 3—28
TRESSL, R. and K.-H. ENGEL, Formation and analysis of optically active aroma compounds, in P. SCHREIER, Analysis of volatiles, W. de Gruyter Berlin/New York 1984, p. 323—342
TRESSL, R. M. HOLZER and M. APETZ, Biogenesis of volatiles in fruit and vegetables, Proc. Internat. Symposium on Aroma Research Zeist/Netherlands 1975, Centre for Agric. Publish. and Doc (Pudoc) Wageningen 1975, p. 51—52
TRESSL, R., T. KOSSA and M. HOLZER, Bildung flüchtiger Aromastoffe durch Maillard-Reaktionen, in F. DRAWERT, Geruch- und Geschmacksstoffe, Internat. Symposium der Fa. Haarmann & Reimer Bad Pyrmont 1974, Verlag Hans Carl Nürnberg 1975, p. 33—47
TURK, A., R. M. SMOCK and T. I. TAYLOR, Mass and infrared spectra of apple vapors, Food Technol. **5**, 58—63 (1951)
URBACH, G., Thin layer chromatography of aliphatic 2.4-dinitrophenylhydrazones, J. Chromatogr. **12**, 196—207 (1963)
URBANYI, GY., Investigation into the interaction of different properties in the course of sensory evaluation. III. Interaction between colour, texture and flavour, Acta Aliment. **13**, 39—46 (1984)
USSEGLIO-TOMASSET, L., I costituento aromatici delle uve, Riv. Viticolt. **22**, 223—242 (1969)
VAROQUAUX, P. and C. AVISSE, Les problemes de flaveur liés aux constituants lipidiques et à leur oxydation dans les légumes, Rev. franc. Corps gras **22**, 373—378 (1975)
VETTER, K., Bedeutung und Einsatzmöglichkeiten kalorienreduzierter Nahrungsmittel, Ber. 5. Tagung Internat. Probleme moderner Getreideverarb. u. Getreidechemie, Eigenverlag Inst. Getreideverarb. Bergholz-Rehbrücke 1973, p. 17—26
VIRTANEN, A. I., Studies on organic sulphur compounds and other labile substances in plants, Phytochemistry **4**, 207—228 (1965)
VITZTHUM, O. G., Chemie und Bearbeitung des Kaffees, in O. EICHLER, Kaffee und Coffein, Springer-Verlag Berlin, Heidelberg, New York 1976, p. 3—64
VODOZ, C. A. Some aspects of the toxicity of flavouring substances, Bibl. Nutritio et Dieta, Verlag Karger Basel/New York 1967, vol. 9, p. 87—104
VODOZ, C. A., Arômes et toxicologie, Mitt. Gebiete Lebensmittelunters. u. Hyg. **60**, 434—444 (1969)
VORBECK, M. L., L. R. MATTICK, F. A. LEE and C. S. PEDERSON, Volatile flavor of sauerkraut. Gas chromatographic identification of a volatile acidic off-odor, J. Food Sci. **26**, 569—572 (1961)
VUATAZ, L. and D. REYMOND, Mathematical treatment of GC data: application to tea quality evaluation. J. chromatogr. Sci. **9**, 168—172 (1971)
WALTKING, A. E., Progress report of the AOCS flavor nomenclature and standards committee, J. Amer. Oil Chemist's Soc. **59** 116A—120A (1982)
WALTKING, A. E. and A. G. GOETZ, Instrumental determination of flavor stability of fatty foods and its correlation with sensory flavor responses, Crit. Rev. Food Sci. **19**, 99—132 (1983)
WEHRLI, A. and E. KOVATS, Gaschromatographische Charakterisierung organischer Verbindungen. 3. Mitt. Berechnung der Retentionsindices aliphatischer, alicyclischer und aromatischer Verbindungen, Helv. chim. Acta **42**, 2709—2736 (1959)
WEURMAN, C., Lists of volatile compounds in food, Central Institute of Food and Nutrition Zeist/Netherlands 1963, Report no. 1687
WEURMAN, C., Isolation and concentration of volatiles in food odor research, J. agric. and Food Chem. **17**, 370—384 (1969)
WEURMAN, C., Vergleich zweier Methoden für die Messung von Gerüchen, VDI-Ber. **226**, 135—139 (1975)

WEURMAN, C., P. J. GROENEN and L. J. VAN GEMERT, Experiments on 'High vacuum transfer' in food odour research, Nahrung 14, 607—616 (1970)

WHITFIELD, F. B., The chemistry of food acceptance, CSIRO Food Res. Quarterly 42, 52—57 (1982)

WHITFIELD, F. B., Some flavours which industry could well do without case studies of industrial problems, CSIRO Food Res. Quarterly 43, 96—106 (1983)

WIENTJES, A. G., The influence of sugar concentrations on the vapor pressure of food odor volatiles in aqueous solutions, J. Food Sci. 33, 1—2 (1968)

WIGHTWICK, I. M., The future of flavour research and development in practice, Food Technol. New Zealand 8, 8, 8—15 (1973)

VAN WIJK, R., The use of poly-para-2.6-diphenylene-oxide as a porous polymer in gas chromatography, J. chromatogr. Sci. 8, 418—422 (1970)

WILLHALM, B. Einige Aspekte der niederauflösenden Massenspektrometrie bei der Untersuchung natürlicher Aromen, Mitt. Gebiete Lebensmitteluntersuch. u. Hyg. 60, 450—452 (1969)

WILLIAMS, A. A., The development of a vocabulary and profile assessment method for evaluating the flavour contribution of cider and perry aroma constituents, J. Sci. Food and Agric. 26, 567—582 (1975)

WILLIAMS, A. A., The aroma components of cider apples and fermented ciders, in F. DRAWERT, Geruch- und Geschmackstoffe, Internat. Symposium Fa. Haarmann & Reimer, Bad Pyrmont 1974, Verlag Hans Carl Nürnberg 1975, p. 141—151

WILLIAMS, P. J., C. R. STRAUSS, B. WILSON and R. A. MASSEY-WESTROPP, Use of C_{18} reversed-phase liquid chromatography for the isolation of monoterpene glycosides and nor-isoprenoid precursors from grape juices and wines, J. chromatogr. Sci. 21, 2013—2020 (1982)

WILSON, R. A. and I. KATZ, Synthetic meat flavour, Flavour-Ind. 5, 1—7 (1974)

WILSON, C. W. III. and P. E. SHAW, Importance of thymol, methyl-N-methyl-anthranilate and monoterpene hydrocarbons to the aroma and flavor of mandarine cold-pressed oils, J. agric. Food Chem. 29, 494—496 (1981)

WINTER, M., F. GAUTSCHI, I. FLAMENT, B. WILLHALM and M. STOLL, Die flüchtigen Aromastoffe des Kaffees, in J. Solms und H. Neukom, Aroma- und Geschmacksstoffe in Lebensmitteln, Forster-Verlag Zürich 1967, p. 182—183

WITTES, J. and A. TURK, The selection of judges for odor discrimination panels, in American Society for Testing and Materials, Correlation of subjective-objective methods in the study of odors and taste, no. 440, Philadelphia 1968, p. 49—70

WIYAND, E., W. NIEMANN, M. KOHLER and C. GEBHARDT, Verfahren zur Verminderung oder Beseitigung des sauren Geschmacks von Erzeugnissen, die normalerweise einen sauren Geschmacks haben, DBP 1940083 (26. 02. 1970)

WÖLM, G. and M. ROTHE, Verkürzte Teigbereitung bei Weißbrot und ihr Einfluß auf das Aroma. 3. Mitt. Einfluß der Hefemenge, Nahrung 18, 165—170 (1974)

WÖLM, G., L. TUNGER and M. ROTHE, Verkürzte Teigbereitung bei Weißbrot und ihr Einfluß auf das Aroma. 2. Mitt. Alkanole als Aroma-Index, Nahrung 18, 157—164 (1974)

WOSKOW, M. H., Selectivity in flavour modification by 5'-ribonucleotides, Food Technol. 23, 1364—1369 (1969)

WRIGHT, R. H., Odor and chemical constitution, Nature [London] 173, 831 (1954)

WRIGHT, R. H., Musk odour and far infrared vibration frequencies, Nature [London] 224, 1033—1035 (1969)

WRIGHT, R. H. and A. ROBSON, Basis of odour specifity: homologues of benzaldehyde and nitrobenzene, Nature [London] 222, 290—292, (1969)

WUCHERPFENNIG, K. and G. BRETTHAUER, Vergleich einiger gas-chromatographischer Untersuchungsmethoden für Aromastoffe von Getränken sowie Beschreibung einer „Säulenextraktion" für Aromastoffe aus Getränken, Z. anal. Chemie 228, 342—360 (1967)

WUCHERPFENNIG, K. and G. BRETTHAUER, Über die Bildung von flüchtigen Aromastoffen in Apfelwein in Abhängigkeit von der Mostverarbeitung sowie von der verwendeten Heferasse, Dtsch. Lebensmittel-Rdsch. 64, 63—67 (1968)

YOUNG, L. L., R. E. BARGMANN and J. J. POWERS, Correlation between gas-chromatographic patterns and flavour evaluation of chemical mixtures and of cola beverages, J. Food Sci. 35, 219—223 (1970)

ZAUSCH, G. T., Repräsentanz von sensorischen Eindrücken durch einzelne einheitliche Riechstoffe als Standard-Bezugssubstanzen, Riechstoffe, Aromen, Körperpflegemittel 20/21, 497—484 (1970/71)

ZAUSCH, G. T., Zur Klassifikation der Gerüche, Mitt.-Bl. GDCh-Fachgruppe Lebensmittelchemie u. gerichtl. Chemie 25, 285—290 (1971)

ZAUSCH, G. T., Stand der Normungsarbeiten im Bereich der Sensorik, DIN-Mitt. 54, 594—596 (1975)

ZEE, J. A., R. E. SIMARD, L. CARBILLET, C. LEBEL and E. LIBER, Comparative composition of fusel oils in brandies made from six grape varieties and their relationship with sensory analysis, Lebensmittel-Wiss. u. -Technol. 17, 54—59 (1984)

ZIEGLEDER, G., Gaschromatographische Röstgradbestimmung von Kakao über methylierte Pyrazine, Dtsch. Lebensmittel-Rdsch. 78, 77—81 (1982)

ZLATKIS, A., Volatile metabolites in normal and pathological samples of biologic origin, Ber. 4. Internat. Symposium on Advances and Utilisation of Chromatography Bratislava 1973

ZOOK, K. and C. WESSMAN, The selection and use of judges for descriptive panels, Food Technol. 31, 11, 56—61 (1977)

ZWAARDEMAKER, H., Olfactometry, Perfumery Essent. Oil Record 12, 308—310 (1921)

Appendix

BELITZ, H.-D. and H. WIESER, Bitter compounds: occurrence and structure-activity relationships, Food Rev. Internat. **1**, 271—354 (1985)

GATFIELD, I. L., Generation of flavour and aroma components by microbial fermentation and enzyme engineering technology, in H. PARLIMENT and R. CROTEAU, Biogeneration of aromas, ACS Symposium Series **317**, Washington 1986, p. 310—322

HOPPE, K., System analysis of the sweetness receptor, Studia biophysica **97**, 97—102 (1983)

HOPPE, K. and B. GASSMANN, Vergleichsmaßstab zur Süßungsintensität von 16 Süßungsmitteln, Lebensmittelind. **32**, 227—231 (1985)

LEE, K.-C. M., H. SHI, A.-S. HUANG, J. T. CARLIN, C.-T. HO and S. S. CHANG, Production of a Romano cheese flavor by enzymic modification of butterfat, in H. PARLIMENT and R. CROTEAU, Biogeneration of aromas, ACS Symposium Series **317**, Washington 1986, p. 370—378

PETRO-TURZA, M., I. SZARFÖLDI-SZALMA, E. MADARASSY-MERSICH, Gy. TELEKY-VAMOSSY and K. FÜZESI-KARDOS, Correlation between chemical composition and sensory quality of natural apple aroma condensates, Nahrung **30**, 765—774 (1986)

POKORNY, J. and J. DAVIDEK, Application of hedonic sensory profiles for the characterization of food quality, Nahrung **30**, 757—763 (1986)

REGLERO, G., T. HERRAIZ and M. D. CABEZUDO, Mixed micropacked columns designed for selective separation of fermentation products, Chromatographia **22**, 358—362 (1986)

ROTHE, M., H. RUTTLOFF, H. HERRMANN and W. ENGST, Problems of technical production and characterization of Blue cheese aroma concentrate, Nahrung **30**, 791—797 (1986)

SCHRÖDTER, R., J. SCHLIEMANN and G. WÖLM, Lipid oxidation products and their importance in the formation of meat aroma, Nahrung **30**, 799—808 (1986)

WHITAKER, R. J., G. C. HOBBIB and L. A. STEWARD, Production of secondary metabolites in plant cell cultures, in T. H. PARLIMENT and R. CROTEAU, Biogeneration of aromas, ACS Symposium Series **317**, Washington 1986, p. 347—362

Bibliography of actual literature to the topic

ADDA, J., Progress in flavour research 1984, Proceedings 4th WEURMAN Flavour Research Symposium Dourdan/France, Elsevier Science Publishers, Amsterdam 1985

AMERINE, M. A., R. M. PANGBORN and E. B. ROESSLER, Principles of sensory evaluation of food, Academic Press, New York 1965

BARYLKO-PIKIELNA, N., Zarys analizy sensoricznej zywnosci, Wydawnictwa Naukowo-Techniczne Warszawa 1975

BERGER, R. G., S. NITZ and P. SCHREIER, Topics in flavour research, Proceedings Internat. Conference Freising-Weihenstephan, Verlag H. Eichhorn, D-8051 Marzling-Hangenham 25, 1985

CHARALAMBOUS, G. and G. INGLETT, Instrumental analysis of foods — recent progress, Vol. 1, Academie Press, New York/London 1983

CROTEAU, R., Fragrance and flavor substances, 2nd Internat. Haarmann u. Reimer Symposium, New York 1979, Verlag Doris u. Peter Siegfried, D-3017 Pattensen 1, 1980

DRAWERT, F., Geruch- und Geschmackstoffe, Internat. Symposium 100 Jahre Fa. Haarmann u. Reimer Holzminden, Bad Pyrmont 1974, Verlag Hans Carl Nürnberg 1975

FURIA, T. E., N. BELLANCA and G. FENAROLI, Fenaroli's handbook of flavor ingredients, Vol. I and II, CRC Press, Cleveland/Ohio 1975

JELLINEK, G., Sensorische Lebensmittel-Prüfung. Lehrbuch für die Praxis, Verlag Doris u. Peter Siegfried, D-3017 Pattensen 1, 1981

JENNINGS, W. and T. SHIBAMOTO, Qualitative analysis of flavor and fragrance volatiles by glass capillary gas chromatography, Academic Press, New York/London 1980

KAWAMURA, Y and M. R. KARE, Umami: a basic taste, Marcel Dekker, New York/Basel 1987

LAND, D. G. and H. E. NURSTEN, Progress in flavour research, Proceedings of the 2nd WEURMAN Symposium, Norwich 1978, Appl. Sci. Publishers Ltd. London 1979

MAARSE, H. and R. BELZ, Handbuch der Aromaforschung, Isolation, separation and identification of volatile compounds in aroma research, Akademie-Verlag Berlin 1981

MAARSE, H. and P. J. GROENEN, Aroma research, Proceedings of the International Symposium on Aroma Research, Zeist 1975, (1st WEURMAN Symposium), Centre Agric. Publish. and Doc., Wageningen 1975

MAARSE, H. and C. A. VISSCHER, Volatile compounds in food. Quantitative data, Vol. 1 (1982), Vol. 2 (1983), Vol. 3 (1984), Vol. 4 (1985), Div. Nutrit. and Food Res. TNO, Inst. CIVO-Analysis TNO, 3700 Zeist AJ

MORTON, I. D. and A. J. MACLEOD, Food flavours. Part A. Introduction, Elsevier Scientific Publish. Co., Amsterdam/Oxford/New York 1982

NYKÄNEN, L. and H. SUOMALAINEN, Handbuch der Aromaforschung, Aroma of beer, wine and distilled alcoholic beverages, Akademie-Verlag Berlin 1983

NYKÄNEN, L. and P. LEHTONEN, Flavour research of alcoholic beverages. Instrumental and sensory analysis, Proceedings of the ALKO Symposium on flavour research of alcoholic beverages, Foundat. Biotechn. and Ind. Fermentat. Res., Vol. 3 (1984)

OHLOFF, G. and A. F. THOMAS, Gustation and olfaction, Internat. Symposium Firmenich Geneva 1970, Food Sci. and Technol. Series, Academic Press, London/New York 1971

PARLIMENT, T. H. and R. CROTEAU, Biogeneration of aromas, ACS Symposium Series **317**, Amer. Chem. Soc., Washington 1986

ROTHE, M., Handbuch der Aromaforschung, Aroma von Brot, Akademie-Verlag Berlin 1974

SCHREIER, P., Flavour '81, Proceedings of the 3rd WEURMAN Symposium, München 1981, Walter de Gruyter Berlin/New York 1981

SCHREIER, P., Analysis of volatiles, Walter de Gruyter Berlin/New York 1984

SCHREIER, P., Chromatographic studies of biogenesis of plant volatiles, Dr. Alfred Hüthig-Verlag Heidelberg/Basel/New York 1984

SCHULTZ, H. W., E. A. DAY and L. M. LIBBEY, Symposium on foods: The chemistry and physiology of flavors, The AVI Publish. Co., Westport/Conn. 1967

SOLMS, J. and H. NEUKOM, Aroma- und Geschmacksstoffe in Lebensmitteln, Fortbildungskurs Eidgenöss. Techn. Hochschule Zürich 1967, Forster-Verlag Zürich 1967

VAN STRATEN, S., H. MAARSE and C. A. VISSCHER, Volatile compounds in food. Qualitative data, Div. Nutrit. and Food Res. TNO, Inst. CIVO-Analysis TNO, 3700 Zeist 1981

TERANISHI, R., R. A. FLATH and H. SUGISAWA, Flavour research — recent advances, Marcel Dekker, New York/Basel 1981

THEIMER, E. T., Fragrance chemistry. The science of the sense of smell, Academic Press, New York/London 1982

ZIEGLER, E., Die natürlichen und künstlichen Aromen, Dr. Alfred Hüthig-Verlag, Heidelberg 1982

Author index

Subject index

GPSR Compliance

The European Union's (EU) General Product Safety Regulation (GPSR) is a set of rules that requires consumer products to be safe and our obligations to ensure this.

If you have any concerns about our products, you can contact us on ProductSafety@springernature.com

In case Publisher is established outside the EU, the EU authorized representative is:

Springer Nature Customer Service Center GmbH
Europaplatz 3
69115 Heidelberg, Germany

Batch number: 10607330

Printed by Printforce, the Netherlands